Green Chemistry in Practice

Advances in Green and Sustainable Chemistry

Green Chemistry in Practice

Greener Material and Chemical Innovation
through Collaboration

Thomas A. McKeag
Senior Advisor and former Executive Director,
Berkeley Center for Green Chemistry, University of California,
Berkeley, United States

Series Editor

Béla Török

Timothy Dransfield

ELSEVIER

Elsevier
Radarweg 29, PO Box 211, 1000 AE Amsterdam, Netherlands
The Boulevard, Langford Lane, Kidlington, Oxford OX5 1GB, United Kingdom
50 Hampshire Street, 5th Floor, Cambridge, MA 02139, United States

Notices
Knowledge and best practice in this field are constantly changing. As new research and
experience broaden our understanding, changes in research methods, professional
practices, or medical treatment may become necessary.

Practitioners and researchers must always rely on their own experience and knowledge in
evaluating and using any information, methods, compounds, or experiments described
herein. In using such information or methods they should be mindful of their own safety
and the safety of others, including parties for whom they have a professional responsibility.

To the fullest extent of the law, neither the Publisher nor the authors, contributors, or
editors, assume any liability for any injury and/or damage to persons or property as a matter
of products liability, negligence or otherwise, or from any use or operation of any methods,
products, instructions, or ideas contained in the material herein.

ISBN: 978-0-12-819674-8

For information on all Elsevier publications visit our website at
https://www.elsevier.com/books-and-journals

Publisher: Susan Dennis
Acquisitions Editor: Charles Bath
Editorial Project Manager: Emerald Li
Production Project Manager: Kumar Anbazhagan
Cover Designer: Mark Rogers

Typeset by TNQ Technologies

Contents

Foreword

The history of green chemistry has been characterized by a high degree of aspirational thinking. Because of this, the application of green chemistry has been highly variable and its practice somewhat uncertain. This is especially true when it comes to educating students and conducting academic research on green chemistry, where published principles and approaches, variously described and vigorously defended, inspire many to study and work on green chemistry. Usually, success in study or research is claimed based on the merits of a one or two principles or attributes that are considered "green" despite the fact that a more holistic review would reveal that "greenness" remains elusive. The beauty of this book is in describing how green chemistry is being taught and practiced in this milieu of uncertainty, clearly illustrating through case studies the challenges and trade-offs that one encounters in actual practice.

This book presents not only the road traveled by the Berkeley Center for Green Chemistry (BCGC) over the course of a decade, but also how others might travel it with fewer bumps and scrapes. Chapter 2 frames the overall approach and philosophy of the Center as a prelude to the case study chapters that follow. The Center has been driven by its close association with the UC Berkeley School of Public Health to focus on the many human health hazards that are associated with chemicals throughout their life cycle. It has also sought strong partnerships with industry to confirm in students the relevance and imperative for green chemistry approaches in design, manufacture, use, and end-of-life considerations for new products. The Center embodies key attributes of green chemistry that are implicit and essential; multidisciplinary problem solving, design-thinking, innovation, environment, safety and health orientation, and partnerships.

The reader will find a wide range of green chemistry challenge examples in this book, across several industry sectors and demanding different associations and solutions. Design contradictions abound and there is rarely one magic solution to these complex problems. Antimicrobials, for example, are important components in contemporary personal care, household, industrial, and institutional consumer products. Chapter 3 is a case study describing BCGC/USDA efforts to replace commonly used preservatives in personal care and household products with less hazardous chemicals. The chapter nicely illustrates the dilemma and design challenges facing manufacturers who need preservatives to ensure product integrity and shelf life while not causing human safety concerns or environmental impacts upon release. While the BCGC/

USDA effort resulted in a patent, a long road to industrial implementation and widespread commercial use remains.

It is generally not appreciated by most consumers that textiles cause a variety of significant human health and environmental impacts throughout their life cycle. After all, most of us just want something that looks good on us for as long as we want to wear it, keeps us warm or cool and is durable enough to last; we don't want to be bothered thinking about where the component parts originate, how and where they're put together, and what happens to them when we donate them to charity. Chapter 4 contains a great case study on efforts to remove a few toxic compounds used in durable water and wrinkle resistance. As is evident from this chapter, the properties we desire currently require the use of hazardous materials, and it isn't easy to replace them.

Collaborations are key to the successes documented here and are often changing over the course of a project. These shifting alliances can bring unexpected innovations. Many industrial sectors have been actively exploring and implementing the use of additive manufacturing technologies over the past 5–10 years. These technologies hold great promise for creating complex metal and plastic parts without the use of expensive molds or extrusion apparatus while potentially removing a significant amount of waste. Chapter 5 contains a case study of finding safer alternatives to toxic acrylate and methacrylate cross-linkers used in stereolithographic photopolymerization. BCGC was able produce several innovations as its partners changed over time, including identifying potential bio-based alternatives and developing a new scorecard approach to chemical hazard assessment. As with most green chemistry, finding less hazardous chemical alternatives that retain the desired function and performance is an iterative and often arduous undertaking.

Throughout this book, and in keeping with BCGC's mission and philosophy, solving pressing societal issues drives research, and this research ranges over geography as well as industry sectors and disciplines. This is well illustrated in Chapter 6. Access to plentiful sources of low-cost energy has given rise to a level of ease and comfort to humans at a scale that is sometimes difficult to comprehend. While fossil carbon fuels have enabled society to extract the sun's energy, this has come at the cost of a range of significant environmental, and increasingly, human health impacts. And, our demand for more energy is insatiable as we move toward greater digitization and automation in modern society. This chapter provides a case study that illustrates the difficult balance between energy demand and how that energy is delivered. Coal in many parts of the world still represents a significant contribution to the electrical grids that power nations' everyday needs. A BCGC research fellow, faculty, and staff looked at the increase in trace metal emissions a new coal-fired power plant in Kosovo would create and then translated these data into human health costs integrated into a cost comparison of coal versus renewable and decentralized power production. These data were used to successfully block World Bank funding guarantees for building a new plant. This is another great example of how a systems-level view of a question can lead to decisions that are protective of human health and the environment.

I commend this book to you for the thoughtful way BCGC has undertaken the practice of green chemistry. It illustrates the profound complexity that comes with decision-making in imagining and producing a more sustainable product. It provides great insight into how to structure learning environments that cultivate responsible and ethical chemists, engineers, and scientists who are equipped to deliver the kinds of sustainable solutions to the challenges facing the world. It's worth your time to read, study, and imagine a different way to make better, safer, chemicals the world desperately needs.

Dr. David Constable
Former Director, Green Chemistry Institute,
American Chemical Society
Wilmington, DE, United States

Acknowledgments

Collaboration, across subject disciplines, market sectors, and professions, is a major theme of this book, and the results offered here would not have been possible without the cooperative efforts of dozens of individuals and organizations. Similarly, whatever achievements are recounted here would not have been possible without decades of scientific progress made by dedicated researchers in the fields of chemistry, engineering, toxicology, and public health.

I want to acknowledge with heartfelt gratitude the following people: for help in the writing of this book, I wish to especially thank Dr. David Faulkner who, as a postdoc at the Berkeley Center for Green Chemistry, researched and wrote the first rough drafts of the case studies; Drs. Billy Hart-Cooper and Kaj Johnson, who reviewed and revised the chapter on their work in reversible-bond antimicrobials; Dr. Rachel Scholes who contributed the written results of the Greener Solutions class of 2019; and Dr. Ann Blake, who shared her deep knowledge of the landscape of green chemistry practice which helped shape the first chapter. The lead participants in our case studies were kind enough to answer our questions during the writing of the first rough drafts. We took pains to mention all of the participants in each chapter because this book is their story and we are so very grateful for their work. Finally, and particularly, I would like to thank my most able project managers at the Berkeley Center for Green Chemisty, Dr. Tala Daya, and (soon to be Dr.) Kim Hazard. If I have forgotten anyone, I apologize.

This book was also built on the solid foundation of tireless work done by the two people who were most instrumental in founding our center, creating and leading the Greener Solutions course, and advocating so adroitly for safer chemistry: Meg Schwarzman MD and Dr. Marty Mulvihill. Their generous advice and support over the years goes well beyond the making of this book. Our program and the Greener Solutions course have been buttressed by dozens of expert volunteers; teaching such an interdisciplinary course would not have been possible without them. I would especially like to thank attorney Claudia Polsky, Dr. Heather Buckley, Dr. Justin Bours, Dr. Akos Kokai, Dr. Larry Weiss, Dr. Tom McKone, Dr. Noah Kittner, and Dr. Mark Dorfman. Over two dozen organizations have partnered with us in addressing chemical and material challenges in the Greener Solutions course and at BCGC; we are grateful for their faith and support. Finally, the results written about in these pages were driven by graduate student power, and each year their inventiveness, capacity for useful work, and spirit of caring inspires me to breathlessness.

Any errors you find in this book are my own and not due to any of the excellent help I have had on this very satisfying job of sharing with you the splendid work of our students, faculty, and associates.

Finally, I want to acknowledge my own personal collaboration with my loving wife, Susan; without her very little work, or joy, would be possible.

Chapter 1

Introduction to Green Chemistry in practice

Green Chemistry

It has been only 2 decades since Green Chemistry was defined as a discreet professional pursuit and methodology in the seminal publication *Green Chemistry: Theory and Practice* by Dr. Paul Anastas and Dr. John Warner. They defined Green Chemistry as the "utilization of a set of principles that reduces or eliminates the use or generation of hazardous substances in the design, manufacture, and applications of chemical products." Anastas and Warner listed 12 principles of practice, all aligned with the model of the Hippocratic Oath to, "first, do no harm." Avoidance, conservation, mitigation or elimination, recycling, and innovation are all strategies to change the practice of chemists in order to produce less harmful products. Philosophically and practically, the authors placed an emphasis on not producing harmful products in the first place, rather than trying to limit exposure or impact.

Importantly, Green Chemistry typically requires intervention at the molecular design phase, generally found in the commodity chemical manufacturing and formulation stages of material making. Emphasis is on the reduction of the inherent hazard of a substance, and generally this means searching for alternatives that are not persistent, bioaccumulative, or toxic (PBT). While practitioners may consider wider mitigating factors, like risk reduction, or weighing of societal cost/benefit of a product or process, their problem-solving focus will typically be on chemical hazard. We consider Green Chemistry to be a vital part of sustainability efforts that can be practiced by all sectors across the entire life cycle of products, processes and systems. This introduction, therefore, includes a brief description of our current sustainability challenges, organized activities to address them, and an attempt to link the practice of Green Chemistry to these activities and the goals behind them.

The practice of Green Chemistry has expanded greatly since 2000, and has become part of the working lexicon and methodology within a wide range of sustainability disciplines.

Green Chemistry in Practice. https://doi.org/10.1016/B978-0-12-819674-8.00007-2

Regardless, the Anastas Warner definition of Green Chemistry has held. Moreover, their 12 principles of practice remain a starting point for best practices in the field. They have guided the teaching and research of the UC Berkeley Center for Green Chemistry and its Greener Solutions graduate course that spawned the case studies within this volume. One of the objectives of this book is to outline an approach to chemical hazard assessment, based on these principles, that results in finding innovative and safer alternatives and contributes to the solving of our pressing world problems of climate change, material scarcity, and health degradation.

Principles of Green Chemistry

1. It is better to prevent waste than to treat or clean up waste after it is formed.
2. Synthetic methods should be designed to maximize the incorporation of all materials used in the process into the final product.
3. Wherever practicable, synthetic methodologies should be designed to use and generate substances that possess little or no toxicity to human health and the environment.
4. Chemical products should be designed to preserve efficacy of function while reducing toxicity.
5. The use of auxiliary substances (e.g., solvents, separation agents, etc.) should be made unnecessary wherever possible and innocuous when used.
6. Energy requirements should be recognized for their environmental and economic impacts and should be minimized. Synthetic methods should be conducted at ambient temperature and pressure.
7. A raw material or feedstock should be renewable rather than depleting wherever technically and economically practicable.
8. Reduce derivatives—Unnecessary derivatization (blocking group, protection/deprotection, temporary modification) should be avoided whenever possible.
9. Catalytic reagents (as selective as possible) are superior to stoichiometric reagents.
10. Chemical products should be designed so that at the end of their function they do not persist in the environment and break down into innocuous degradation products.
11. Analytical methodologies need to be further developed to allow for real-time, in-process monitoring and control prior to the formation of hazardous substances.
12. Substances and the form of a substance used in a chemical process should be chosen to minimize potential for chemical accidents, including releases, explosions, and fires.

Beyond the intention "to do no harm" the field has necessarily grown and expanded with the sustainability movement in general and it is widely accepted that the practice of Green Chemistry is also an opportunity to provide novel efficiencies within the life cycle of products, a regeneration of material resources, and innovation in design. Indeed, molecular technology is one of the faster growing sectors of the 21st century economy, and the devising of safer and more efficient chemical reactions that result in new products and processes will contribute significantly to meeting the pressing sustainability goals of the next 50 years.

Green Chemistry, the making and using of more benign chemicals within our products, processes and systems, remains a relatively small part of the over five trillion dollar overall global chemical market that had been estimated in 2017, and is projected to double by 2030. Pike Research, a market research and consulting company, now part of Navigant Research, had projected a growth of market opportunity for green chemistry of $98.5 billion by 2020, and a projected market growth since 2011 of 35 times. Most of this growth projection was in polymer production. The drivers for this growth spring mainly from the desire to minimize waste, replace existing chemicals with less toxic alternatives (from both a market pull and regulatory push) and to source renewable feedstocks that do not rely on petrochemicals.

Challenges to Green Chemistry adoption remain. T. Fennelly and Associates, in a 2015, report to the Green Chemistry and Commerce Council, *Advancing Green Chemistry: Barriers to Adoption and Ways to Accelerate Green Chemistry Adoption*, outlined nine factors that had slowed or impeded adoption: lack of agreement on a working (quantifiable) definition; supply chain complexity; incumbency of traditional materials; confusion about materials; fear of the risk of change; the dominance of price/performance as the determinant of decision-making; lack of demand to justify supply changes; intellectual property protection and subsequent lack of transparency; lack of access to new technology and innovation.

Sustainability context

The major global trends of climate change, environmental degradation, population growth, urbanization, and the pressure placed on people to obtain the basic necessities of life such as food, energy and water, are all factors that will continue to affect our decisions about which chemicals we make and how we use them. New chemical formulation, production and use have the potential to improve lives by saving energy and materials, but can also come with significant human health and environmental costs. Similarly, existing chemical stockpiles and use have created a harmful legacy as well as an improved way of life, and this legacy requires innovative chemical problem solving as well as discovery of more benign alternatives.

Climate change

In its September 2019 report the World Meterological Organization concluded that the 5 year period between 2014 and 2019 was the warmest on record. In this same year, emissions of carbon dioxide, a major contributor to global warming upon release to the atmosphere, were at an all-time high. Projections are that average global temperatures will be 3°C higher than the mid-19th century (preindustrial age). Current average global temperature is 1.1°C higher. The world is getting hotter and at a faster pace.

The higher temperatures will have an increasing impact on people, with the number needing humanitarian aid likely to double by 2050, caused by phenomena like drought caused famine, scarcity caused conflict and rising seas. A report from 13 United States federal agencies in 2018 warned that failing to contain warming trends could reduce the U.S. economy by as much as 10% by century's end.

While trends toward a hotter planet and more severe impacts appear to be accelerating, more decisive action on the part of world governments also presents the chance for increased mitigation and adaptation.

In 2018 the United Nations Intergovernmental Panel on Climate Change (IPCC) released the special report "Global Warming of 1.5°C" on the impacts of global warming of 1.5°C above preindustrial levels and related global greenhouse gas emission "pathways". This had been requested by the signatories (Conference of Parties (COP)) upon the adoption of the Paris Agreement at the United Nations Framework Convention on Climate Change (UNFCCC) at its 21st Session in Paris, France (30 November to 11 December 2015)).

In the special report the IPCC was to provide analysis of the impacts of global warming of 1.5°C above preindustrial levels and related global greenhouse gas emission pathways for the purposes of strengthening the global response to the threat of climate change, sustainable development, and efforts to eradicate poverty. The authors found, with high confidence, that human-induced warming had reached approximately 1°C above preindustrial levels in 2017, and that this type of warming was increasing at approximately 0.2°C per decade.

Limiting warming to these levels of 1.5°C would significantly reduce the risks of scarcity, ill health, conflicts and the occurrence of floods, droughts, extreme heat and tropical storms. It would also reduce the alarming trends of biodiversity loss and sea level rise. Achieving this would require both mitigation and adaptation to reach zero emissions by 2050, and the continued removal of CO_2 for the rest of the century. The IPCC targeted methane, black carbon, and other superpollutants as critical to the drastic emission reductions now needed. These reductions would have to occur within broader programs to:

Decrease across-the-board energy demand.
Lower emissions from the energy supply sector.
Actively remove carbon dioxide from the atmosphere.
Fully decarbonize the electricity sector by mid-century.
Ensure renewables are the world's dominant energy source by 2050.
Balance land-use between sustainable agriculture practices, bioenergy production, and carbon storage.

Green chemistry has a part to play in the two main areas of response to the climate crisis, mitigation, and adaptation. Developing alternatives to fossil fuel-based chemicals, increasing the efficiency of chemical reactions in manufacturing, synthesizing new materials to avoid material and energy costs and associated emissions are three ways that green chemistry can contribute to reducing the rise of global temperatures. One example is in the substitution of bio-based for petroleum-based industrial feedstocks. Toensmeier and Blake, in *Industrial Perennial Crops for a Post Petroleum Economy,* lay out various ways that perennial crops can "… provide biomass, starch, sugar, oil, hydrocarbons, fiber, and other products. Biomass feedstocks can replace a variety of petroleum-based chemicals currently used to manufacture solvents, resins stabilizers, dispersants, binders, and fillers. Starch feedstocks can be used to manufacture solvents, paints, glues, coagulants, flocculants, textile finishing agents, and many other materials. Perennial industrial crop oils can be made into glycerin, soaps, lubricants, surfactants, and surface coatings. Plant-sourced hydrocarbons can be used as feedstocks for the full range of modern industrial chemistry."

Material scarcity and the nexus of food, energy, and water

Material scarcity can be typified in three categories; demand-induced, supply induced and structural. Rising population and income growth will increase demand for certain goods and this will exceed the supply, for example. Conversely, events like drought or war may limit the supply of a material and cause scarcity. Finally, geography or political imbalance may limit the access one part of a population may have to a good; this can also be defined as scarcity. We will limit this discussion to the broad lens of demand trends and the well-documented projections for population and economic growth, particularly of emerging and developing countries.

Food, Energy, and Water are inextricably linked and the development and procurement of each affects the others. Water needs to be moved and stored and treated and distributed; food needs to be raised, protected, harvested, stored and distributed; energy needs to be captured, stored, and distributed. For example, increased food production implies increased energy consumption, water requirements, and material expenditures, as well as emissions, waste and impacts on the physical environment. Agriculture, for example, accounts for

70% of the freshwater withdrawals in the world. Commodity prices are affected by the fluctuation in energy and material costs. For example, oil and gas prices typically influence directly commodity prices in the global market. Energy production requires water for industrial extraction, equipment cooling, or growing bio feedstocks, and agriculture consumes 30% of total energy consumed worldwide.

The United Nations Food and Agriculture Organization (FAO), and several other organizations have championed the strategic approach of insuring the security of people in these three areas of life support. Security in Food, Energy and Water is seen as foundational for sustainability and supporting the broader UN Sustainable Development Goals (SDGs) described below. This approach recognizes the complex and dynamic interrelationships among these sectors, in an effort to address the tradeoffs and synergies that arise in the allocation of increasingly scarce resources. Security is defined as the assurance of availability, access, stability, and utilization.

The world is getting more crowded and this will stretch our ability to feed ourselves. The population of the world is predicted to rise to over nine billion by 2050 with the Food and Agriculture Organization (FAO) of the United Nations (UN) projecting that global food production will need to grow by 60% from 2005/7 levels to support increased demand. Total water withdrawals for irrigation alone are expected to increase by 10% by 2050.

There will be more demand for energy. The U.S. Energy Information Administration (EIA) projects that world energy consumption will grow by nearly 50% between 2018 and 2050. Most of this growth will come from rising demand from economic growth in Asia. In International Energy Outlook 2019, researchers assessed long-term world energy markets for 16 regions of the world, dividing results between members of the Organization for Economic Cooperation and Development (OECD) and nonmembers. Total global primary energy consumption is predicted to reach 900 quadrillion British Thermal Units (BTUs) by 2050, with Asian nations accounting for approximately 44% of this total. The industrial sector, comprising refining, mining, manufacturing, agriculture, and construction, accounted for the largest share of energy consumption of any end-use sector—more than half (http://www.eia.gov/outlooks/ieo/).

Much of the world is now in a water supply crisis. Over two billion people live in countries experiencing high water stress (UN, 2018); and of these 700 million people worldwide could be displaced by intense water scarcity by 2030 (Global Water Institute, 2013). Currently a third of the world's biggest groundwater systems are already in distress (Richey et al., 2015), and nearly half the global population are already living in potential water-scarce areas at least 1 month per year and this could increase to some 4.8–5.7 billion in 2050. About 73% of the affected people live in Asia (69% by 2050) (Burek et al., 2016) (https://img1.wsimg.com/blobby/go/27b53d18-6069-45f7-a1bd-d5a48 bc80322/downloads/1c2meuvon_105010.pdf).

Despite scarcity in basic materials and energy, the world economy will continue to grow and place more demand on them. The world economy, measured in global domestic product (GDP) is projected by the OECD to quadruple by 2060, driven largely by the rise in incomes in China, India and other emerging and developing nations. Average consumption of goods, and therefore material production will rise. Global material use is projected to double from 2011 to 2060, from 79 Gt to 167 Gt; nonmetallic materials like sand, gravel and crushed rock are expected to make up more than half of total material use.

The OECD also predicts that material production will be slightly decoupled from economic growth as the service sector grows as a percentage of the economy. Moreover, recycling will become more efficient than primary extraction and represent a larger share of materials handling. Finally technological and societal innovation will be a factor in reducing material needs for comparable value. Lightweighting of materials in products and the leasing/ buyback of products are examples.

Green chemistry will continue to be an important part of strategies to provide the basic necessities to a growing world population. Whether it is improving the yield of food crops, developing a more efficient water treatment system, or increasing the yield of biofuels, these endeavors, by their nature, will demand a reasonable level of safety for humans and the environment upon which we depend. One example in water treatment is the use by Dibakar Bhattacharyya and colleagues of enzymes and nanostructured membranes to remove organic pollutants from water. Tainted water is permeated with glucose and then passed through two filters, the first loaded with enzymes to produce hydrogen peroxide from the glucose and the second loaded with iron for the decomposition of the hydrogen peroxide to form powerful free radical oxidants that can break down organic pollutants like trichlorophenol.

Environmental pollution and human health

Pollution is the largest environmental cause of disease and premature death in the world today, according to the *Lancet* Commission on Pollution and Health. Diseases caused by pollution were responsible for an estimated nine million premature deaths in 2015—16% of all deaths worldwide. The study points out that this is three times more deaths than from AIDS, tuberculosis and malaria combined. In severely affected countries, this accounts for more than one death in four.

Pollution is typically categorized by where it occurs and we will discuss briefly general pollution in air, water and soil. Ultimately all pollution (and solutions to it), however, has a chemical component, so we believe it is important to distinguish chemical pollution as a critical component of the air, water and soil pollution caused by the by-products of industrial society. Chemical pollution can be described here as the threat or injury to environmental and human health caused by the production and use of inherently hazardous synthetic compounds. These harmful synthetic compounds are the focus of many of the case studies in this volume, and the search for alternatives to them the main work of the Berkeley Center for Green Chemistry.

Chemical pollution is a great, growing, and poorly defined problem. The sheer scale of production and the lack of intelligence on health impacts are daunting. More than 140,000 new chemicals and pesticides have been created since 1950, and the 5,000 largest volume ones produced are now universally disbursed throughout the globe and in virtually all populations. Of these, only half have ever undergone any testing for safety or toxicity. Legacy chemicals known to have caused disease, death and environmental degradation include lead, asbestos, dichlorodiphenyltrichloroethane (DDT), polychlorinated biphenyls (PCBs), and the ozone destroying chlorofluorocarbons. Newer synthetic chemicals and compounds of concern include developmental neurotoxicants, endocrine disruptors, chemical herbicides, novel insecticides, pharmaceutical wastes, and nanomaterials.

Air pollution

We discuss two types of air pollution here, ambient and household, the connection to harmful chemical production, and how the practice of green chemistry can mitigate some of the impacts. Ambient and household combined air pollution are responsible for an estimated seven million premature deaths per year, according to the World Health Organization (WHO) and nine out of 10 people in the world breathe air that is heavily laden with pollutants. Premature deaths are mainly attributed to increased mortality from stroke, heart disease, chronic obstructive pulmonary disease, lung cancer, and acute respiratory infections.

Ambient air pollution is dispersed globally and does not recognize national boundaries, and its cumulative effect on the protective ozone layer of the earth's atmosphere has produced our current climate crisis of increased temperatures and more severe weather fluctuations. Air pollution also kills on a daily basis. It is estimated by the WHO to be responsible for 4.2 million premature deaths every year. The Health Effects Institute estimates that China and India account for about 50% of the air pollution health burden worldwide. In China, the main sources of this pollution are coal burning by industry, power plants and residential heating (40%), followed by transportation and residential biomass burning. In India, residential biomass burning for cooking and heating (25%), followed by coal burning by power plants and industry, were the main sources.

Ambient air pollution is measured by several important constituents of the air that form the basis for the WHO Air Quality Guidelines: particulate matter (PM), ozone (O_3), nitrogen dioxide (NO_2), and sulfur dioxide (SO_2). PM is the most affecting of the ingredients in polluted air and is the most used surrogate for measuring. It comprises a complex mix of solid and liquid particles of organic and inorganic substances in suspension. Its main components are sulfate, nitrates, ammonia, sodium chloride, black carbon, mineral dust, and water.

PM is measured in microns and any particle less than 10 microns can enter, lodge and damage the lungs. Particles of 2.5 microns and less in diameter are the most damaging, however; at this scale the particles are able to cross the epithelium of the lungs and enter the blood stream where they can contribute to the risk of developing cardiovascular and respiratory diseases, as well as of lung cancer. There is a direct correlation between exposure to high concentrations of PM and mortality and morbidity. As important to note, exposure to PM is damaging even at low concentrations.

Ozone (O_3) in the upper atmosphere serves as a buffer against the harming rays of the sun. When it is at ground level in high concentrations, however, it can be a serious health concern. It is one of the main ingredients of photochemical smog-the reaction of sunlight and pollutants like nitrous oxides and volatile organic compounds (VOCs) coming from a variety of sources like vehicle exhaust and industry emissions. Excessive ozone can cause asthma, reduced lung function and lung disease.

Nitrogen dioxide (NO_2) is the main source of nitrate aerosols and, in the presence of ultraviolet light, a precursor to ozone. Nitrate aerosols are a significant portion of the less than 2.5 micron PM pollutants that can be especially harmful. At high enough concentrations (greater than 200 ug/m^3) NO_2 is a toxic gas that causes inflammation of the airways. Its main sources are combustion engines in vehicles, and power plants. There has been a direct correlation shown between NO_2 exposure and rates of bronchitis symptoms in asthmatic children as well as reduced lung function growth.

Sulfur dioxide (SO_2) is formed from the burning of anything that contains sulfur. This is typically fossil fuels in power plants and the internal combustion engines of vehicles or the smelting of minerals. Human tolerances for SO_2 are low, with even 10-min exposures linked to respiratory inflammation and eye irritation. SO_2 has been shown to aggravate asthma and chronic bronchitis, and increased exposure has been correlated with more hospital admissions for cardiac disease and mortality. Mixed with water, sulfuric dioxide makes sulfuric acid, the main component of acid rain which causes deforestation.

Household air pollution has long been studied as a problem of emerging and developing economies. The use of traditional cooking and heating methods and the dangers to human health of the particulate matter and uncombusted chemicals in smoke have been characterized well. There are an estimated three billion people who cook and heat their homes with biomass, kerosene fuels and coal, and they are subject to the associated health effects. Household air pollution causes noncommunicable diseases including stroke (18%), ischemic heart disease (27%), chronic obstructive pulmonary disease (COPD-20%) and lung cancer (8%). As with most pollution, household air pollution disproportionately affects the poor, the very young and the very old. For example, close to half (45%) of deaths due to pneumonia among children under 5 years of age are caused by PM inhaled from indoor air pollution, most of which is created by cooking and heating with solid carbon fuels or kerosene that are incompletely combusted. Because women do almost all the cooking in these traditional homes, they are also the most susceptible. Nearly 3.8 million people die prematurely from illnesses associated with indoor air pollution.

There is, moreover, a substantial body of evidence of the dangers of indoor air pollution of a different nature. Chemical off-gassing from industrial products like construction materials and furniture, the use and storage of household cleaners and personal care products have created unhealthy indoor conditions in developed countries as well. The US EPA's Office of Research and Development's "Total Exposure Assessment Methodology (TEAM) Study," 1985, found levels of about a dozen common organic pollutants to be 2 to 5 times higher inside homes than outside, regardless of where the homes were located. Constituents of indoor air pollution in developed countries are likely to be bio-aerosols like bacteria, molds and viruses; VOCs, carbon-containing gases given off by a wide range of materials; and other gases or particulates like lead, asbestos, radon, and carbon monoxide.

VOCs have been of particular concern and subject to study and include formaldehyde, benzene, methylene chloride, tetrachloroethylene, heptane, naphthalene, and ethyl acetate. They may be given off from paints, adhesives, solvents, preservatives, cleansers and disinfectants, air fresheners, fuels and automotive products, hobby supplies, and pesticides.

Many of the ingredients in indoor air pollution are highly toxic. Lead is a neurotoxin; formaldehyde, asbestos, benzene and radon are classified as Group 1 Carcinogens by the International Agency for Research on Cancer (IARC) of the UN's WHO; carbon monoxide can be fatal in high concentrations; tetra-chloroethylene and naphthalene are Group 2 Carcinogens.

In addition to the direct human health impact of daily exposure to VOCs, they have also been found to be an increasing segment of the emissions from fossil fuels that are contributing to the climate crisis. One 2018 study by McDonald et al., estimated that VOCs from pesticides, coatings, inks, adhesives, cleaning agents, and personal care products make up half of all VOCs from fossil fuels in industrialized cities.

Water pollution

The Global Burden of Disease (GBD) study done by the *Lancet* in 2015 estimated that 1.8 million deaths per year were attributable to water pollution, including unsafe water sources, unsafe sanitation and inadequate hand washing. Unsafe water sources accounted for 1.3 million deaths, with overlap between the reasons noted. The GBD does not quantify deaths attributable to chemical pollution per se, so it is beyond the scope of this text to differentiate, but we will note some individual case figures.

About 70% of the diseases linked to water pollution are diarrheal, with typhoid and paratyphoid fevers and lower respiratory tract infections making up most of the rest. These diseases affect more than one billion people, most in low-income and middle-income countries and most are the very young, below 5 years of age. The WHO calculated lower respiratory and diarrheal diseases to be the number one and number two causes of death rates overall in low-income countries in 2016.

Water Pollution in particular has a wider health impact as it is highly injurious to the basis of ecosystems, particularly vulnerable nursery and feeding grounds and can collapse the resources that human populations depend on, such as fisheries.

Freshwater and Marine water bodies can become contaminated with hazardous chemicals through both point sources like industrial facilities and municipal wastewater treatment facilities, and nonpoint sources such as fields, roads, and parking lots. Worldwide, more that 80% of municipal and industrial wastewater is released to the environment without adequate treatment.

Agriculture remains the largest source of surface water pollution globally. Farming and food processing generate some 40% of water pollution in higher-income countries and 54% in lower-income countries. Insecticides, herbicides and fungicides represent a significant portion of the hazardous chemicals being passed to waterways. Fertilizers that contain nutrients like nitrogen that overload natural systems cause algal blooms and eutrophication of waterways that kills aquatic life and pollute groundwater that is often the source of community drinking supplies.

Industrial discharge and household wastewater are the two other main generators of water pollution. Industrial discharge may contain heavy metals, solvents, oils and greases, acids, and other aquatic toxicants. Solvents from industry such as trichloroethylene, perchloroethylene, and methylene chloride are examples of a common source of groundwater contamination.

Pharmaceuticals and personal care products from households have become increasingly concerning to public health officials as active ingredients in formulas make their way into natural water systems. Approximately 4,000 active pharmaceutical ingredients are prescribed in prescription and nonprescription drugs and some 100,000 tons produced every year. The flushing of antibiotics and synthetic hormones into waterways has created the growth of antibiotic resistant pathogens, injurious to humans but now unresponsive to traditional treatments.

Approximately 80% of ocean pollution comes from land-based sources, including sewage discharges, litter, and runoff from industrial and agricultural sources. In many developing countries, most of the sewage discharge is completely untreated and can contain heavy metals such as cadmium, mercury, and lead. It can also include persistent organic pollutants (POPs). Familiar examples are PCBs, DDT, and dioxins, but there are many others intentionally used in agriculture, industry and also in commercial products. The amount of plastic released into the oceans is immense and runs the linear scale from macro- to nanosize. The total mass of plastic debris added to the marine environment from 2010 until 2025 is expected to grow by an order of magnitude and may amount to some 100−250 million tons. Most of the microplastic waste comprises clothing and textiles broken down by machine washing (35%) and the grinding of automobile tires on roads (29%).

Soil pollution

Agriculture, mining, manufacturing, and waste disposal are the sources of hazardous chemicals entering the soil. Waste disposal might be solid waste in landfills or surface application of sewage sludge. Mining has the biggest impact on soil health. Zinc, copper, manganese, lead, nitric acid, and their compounds are the main pollutants released to the soil in North America, and mining wastes represent 44% of the reported releases of chemicals in the U.S. Nearly all of this is placed upon the soil. In addition to the heavy metals listed above, wastes from mining often contain arsenic and cyanide produced in the extraction process. Highly acid mine drainage is the largest source of both soil and water pollution from the mining industry. Mercury is still used in small gold mining operations worldwide and this activity produces over 1,000 tons of this pollution per year and represents the largest source of mercury release to the soil.

Pesticides and fertilizers are applied directly to crops and the soil and therefore have the potential for greater local impact to soil health than the settling of airborne toxicants or water runoff. Fertilizer and pesticide use continue to increase globally, while the intensity of applications has decreased, according to the Global Chemicals Outlook II of the UN Environment Program, 2019. More than 20,000 different commercial pesticide products are now marketed globally, and an estimated 5.2 billion pounds produced each year.

Herbicides made up the largest share of the pesticides (around 40%), and the most common herbicides were glyphosate-based. Glyphosates are the largest-volume herbicides in use today, and their use has grown in 2 decades by one order of magnitude, from over 56,000 tons in 1994, to over 825,000 tons in 2014. Glyphosate sales have been linked to the sale of proprietary, genetically engineered crops, particularly soybeans and corn. In the U.S. nearly 90% of these crops have been engineered to resist the glyphosate used to control weeds in the crop rows. The WHO's International Agency for Research on Cancer (IARC) has determined glyphosate to be a "probable human carcinogen."

As in water pollution, microplastics are increasingly being found in soils and one-third of all plastic waste is estimated to end up in freshwater or soils.

Municipal sewage treatment plants distribute microplastics through discharge of waste water and the land application of sewage sludge in agriculture. The Norwegian Institute for Water Research identified sewage sludge as an important source of microplastics in agricultural soils in 2018, and estimated that 110,000−730,000 tons of microplastics are released per year to agricultural soils in Europe and North America.

Persistent substances, such as heavy metals, POPs, and radionuclides present the greatest threats to human health at sites with contaminated soils. Mercury, lead, chromium and cadmium are the metals most commonly encountered worldwide. As in most environmental pollution, the majority of contamination and the resultant health effects occur in low-income and developing countries.

Exposure to heavy metals and toxic chemicals at contaminated sites is widespread. About 61 million people in 49 countries surveyed were exposed to these threats to human health according to the GBD study done by the *Lancet* in 2015. This is an undercount as it surveyed only a fraction of total sites worldwide.

Infrastructure and occupational pollution

Ours is an increasingly urbanized world and pollution within this built environment cannot be ignored in any discussion of human health. Buildings and other structures retain substances for a relatively long time and this includes potentially harmful materials like lead in paint, asbestos in tiles, perfluorinated compounds in framing, furniture, and floor coverings. This is also the environment in which most people in the high and middle income countries work, so exposure to some of these materials is chronic as well as prolonged.

Occupational diseases associated with industrialization have included pneumoconiosis and silicosis associated with coal mining, bladder cancer (dyeworks); leukemia and lymphoma from exposure to benzene; asbestosis, lung cancer and mesothelioma from exposure to asbestos. As economies have changed so have the potential threats to workers and currently there is a wide range of synthetic chemicals used in the manufacture of products that pose a threat to health.

Carcinogens were responsible for 55% of the 0.88 million deaths globally from occupational risk factors in 2015, according to the GBD study, with asbestos responsible for nearly 40% of these occupational carcinogen deaths. Exposure to particulates, gases, and fumes were responsible for the bulk of the remainder of the deaths attributable to occupational risk.

The practice of Green Chemistry can have a significant impact on these three existential challenges of climate change, material scarcity, and environmental pollution and below we discuss this impact within the framework of the Sustainability Goals of the United Nations 2030 Agenda for Sustainable Development. This framework is the broad set of contextual goals used by the

Berkeley Center for Green Chemistry in our strategic planning and organizational performance assessments (Fig. 1.1).

UN Sustainable Development Goals (SDGs)

The 2030 Agenda for Sustainable Development, adopted by all United Nations Member States in 2015, contains 17 Sustainable Development Goals (SDGs) as a shared blueprint for present and future peace and prosperity for the world and its peoples. The goals reflect the belief that human and environmental health must go hand in hand with economic growth, equality, and education (https://sustainabledevelopment.un.org/post2015/transformingourworld).

Sustainable Development Goals

Goal 1. End poverty in all its forms everywhere.

Goal 2. End hunger, achieve food security and improved nutrition, and promote sustainable agriculture.

Goal 3. Ensure healthy lives and promote well-being for all at all ages.

Goal 4. Ensure inclusive and equitable quality education and promote lifelong learning opportunities for all.

Goal 5. Achieve gender equality and empower all women and girls.

Source: Erythropel, Hanno C., et al. "The Green ChemisTREE: 20 years after taking root with the 12 principles." Green Chemistry 20.9 (2018): 1929-1961. Table 3

FIGURE 1.1 Categorizes the hazards presented by the current manufacture, use, and disposal of synthetic chemicals. To be noted here is the wide range of scale of hazard represented, from global concerns like climate change to molecular scale toxicity. *Adapted from Erythropel, H. C., Zimmerman, J. B., de Winter, T. M., Petitjean, L., Melnikov, F., Lam, C. H., ... & Anastas, P. T (2018). The Green ChemisTREE: 20 years after taking root with the 12 principles. Green chemistry, 20(9), 1929–1961.*

Goal 6. Ensure availability and sustainable management of water and sanitation for all.

Goal 7. Ensure access to affordable, reliable, sustainable, and modern energy for all.

Goal 8. Promote sustained, inclusive, and sustainable economic growth, full and productive employment and decent work for all.

Goal 9. Build resilient infrastructure, promote inclusive, and sustainable industrialization and foster innovation.

Goal 10. Reduce inequality within and among countries.

Goal 11. Make cities and human settlements inclusive, safe, resilient, and sustainable.

Goal 12. Ensure sustainable consumption and production patterns.

Goal 13. Take urgent action to combat climate change and its impacts.

Goal 14. Conserve and sustainably use the oceans, seas, and marine resources for sustainable development.

Goal 15. Protect, restore, and promote sustainable use of terrestrial ecosystems, sustainably manage forests, combat desertification, and halt and reverse land degradation and halt biodiversity loss.

Goal 16. Promote peaceful and inclusive societies for sustainable development, provide access to justice for all and build effective, accountable and inclusive institutions at all levels.

Goal 17. Strengthen the means of implementation and revitalize the global partnership for sustainable development.

UNEP's Global Chemicals Outlook Report II, 2019, is a comprehensive assessment of the state of global chemicals management and remaining capability shortfalls in meeting UN Sustainable Development Goals by 2030. It also identifies opportunities to link international chemical management policies within different sectors of the value chain.

Of these 17 goals, the most directly aligned with green chemistry principles as they relate to public health is *Goal Three: Ensure healthy lives and promote well-being for all at all ages.* More specifically, Target 3.9 states the following: *By 2030, substantially reduce the number of deaths and illnesses from hazardous chemicals and air, water and soil pollution and contamination.*

Moreover, as stated by the UN, "… *Sound management of chemicals and waste is a specific target under* SDG 12 *on Sustainable Consumption and Production. Chemicals, waste and air quality are also referred to under* SDG 3, *on Good Health and Well-being;* SDG 6, *on Clean Water and Sanitation;* SDG 7, *on Affordable and Clean Energy;* SDG 11, *on Sustainable Cities and Communities; and* SDG14, *on Life Below Water."*

Target 12.4 states: *By 2020, achieve the environmentally sound management of chemicals and all wastes throughout their life cycle, in accordance with agreed international frameworks, and significantly reduce their release to air, water and soil in order to minimise their adverse impacts on human health and the environment.*

Here are the Sustainable Development Goals. we deem most directly aligned with Green Chemistry:

A. the reduction and mitigation of climate change (Goal 13)
B. the production of food, and energy and the treatment of water (Goals 2, 6, 7)
C. the more efficient use and recycling of materials, and the innovative design and production of new materials (Goal 12)
D. the assurance of the daily health of people by providing safer products, work and living environments, and better nutritional, pharmaceutical and medical care (Goal 3)

The SDGs have been adopted by many organizations from all sectors across the globe. The following four have incorporated the framework into their activities. The International Pollutants Elimination Network (IPEN) advocates for more aggressive action on eliminating toxic chemicals, and has focused on agroecology, chemicals in products, and endocrine-disrupting chemicals, pesticides, lead in paint, women and workers, and zero waste in their Toxic-Free SDGs Campaign. The World Business Council on Sustainable Development (WBCSD) has produced a Chemical Sector Roadmap that addresses all 17 goals and includes case studies. The nonprofit B Lab, which promotes the Benefit Corporation movement, has produced a tool for businesses, the SDG Action Manager, that allows companies to track their progress toward goals that they select. Finally, Clean Production Action (CPA) has incorporated SDGs 3, 6, and 12 into their Chemical Footprint Project (CFP) Framework, a tool for companies to assess and track progress toward eliminating chemicals of concern from their value chains.

While we have outlined the most direct applications of green chemistry alternatives, the practice can have impact over a wide range of endeavors associated with the SDGs.

Chemical policy and regulations

The increasing complexity and interconnectedness of the global economy has had an effect on commerce, but also on the array of regulations promulgated to control its operation and impact. Included are the regulations directly related to human health and the environment as they relate to chemical manufacturing and use. Below we provide a broad outline of these regulations, divided first into the most germane international policy conventions, European Union regulations, U.S. federal regulations, and state regulations practiced in California. In addition we discuss some self-regulatory or cooperative activities in the private sector. It should be noted that there are significant policies and regulations from other sources at different geographical scales that may have a significant global impact, but that are out of the scope of this publication.

Germane activities in the international, US federal, and in the State of California have been included here because of their consideration in most of the case studies pursued in the Greener Solutions program and detailed in later chapters. These activities can be divided into policy, law, standards and resources, and all have an impact on the type of innovative solutions that have been possible.

International

Most international chemical management policy agreements have been accomplished through treaties or agreements under the auspices of the United Nations Environment Program.

Strategic Approach to International Chemicals Management (SAICM)

SAICM is a voluntary agreement among international intergovernmental and nongovernmental organizations. It is administered by the UN Environment Program. The agreement defines a policy framework to foster the sound worldwide management of chemicals. SAICM's overall objective is "… the achievement of the sound management of chemicals throughout their life cycle so that by the year 2020, chemicals are produced and used in ways that minimize significant adverse impacts on the environment and human health." This framework includes guidance on risk assessment, labeling, and disposal, as well as setting up national centers, safety training, and emergency accident training. This initiative was adopted in 2006, and updated in 2009. It expired in 2020, and members met in October of that year for the fifth International Conference on Chemicals Management to extend and expand SAICM through 2030. The Declaration and Strategy of the SACIM cites five objective themes: risk reduction, information, governance, capacity-building and technical cooperation, and illegal international traffic. This is accompanied by a Global Plan of Action for implementation by members.

A trio of international agreements among national governments (and regional economic integration organizations) related to chemicals, the Basel, Rotterdam and Stockholm Conventions (BRS), are overseen by secretariats within the United Nations Environment Program. The three conventions have mechanisms for so-called synergies, including joint sessions and sharing of resources.

Basel Convention

The Basel Convention on the Control of Transboundary Movements of Hazardous Wastes and their Disposal is a global environmental agreement on hazardous and other wastes among governmental parties. It came into force in 1992. The Convention addresses the generation, management, transboundary movements and disposal of hazardous and other wastes, and is organized by a Conference of the Parties (COP), an implementation committee, and an open-ended working group.

Rotterdam Convention

The Rotterdam Convention is an agreement to: promote shared responsibility and cooperative efforts in the international trade of certain hazardous chemicals. Members share information about chemical characteristics, best practices, and relay national decision-making on import and export to all parties The Convention entered into force on February 24, 2004. The Convention creates legally binding obligations for the implementation of the Prior Informed Consent (PIC) procedure and facilitates information exchange for a broad range of hazardous chemicals. The convention maintains a list of 52 pesticide and industrial compounds and formulations in its Annex III. Its Chemical Review Committee reviews notifications and proposals from Parties, and makes recommendations to the Conference of Parties (COP) on the addition of chemicals to Annex III.

Stockholm Convention

The Stockholm Convention is a global treaty to protect human health and the environment from POPs. POPs, or persistent organic pollutants, are organic compounds that are resistant to environmental degradation through chemical, biological, and photolytic processes. It entered into force, on May 17, 2004, targeting 12 initial chemicals of concern and has subsequently added 16 more. The convention classifies chemicals of concern for three types of response commitment: elimination of the chemical altogether (Annex A), reduction of the chemical (Annex B), and reduction of the unintentional release of the chemical (Annex C). Examples, in order, include Hexachlorobenzene (HCB) initially used as a fungicide to treat crops, Perfluorooctane sulfonic acid, a pesticide and industrial chemical, and Polychlorinated biphenyls (PCBs), unintentional by-products of several industrial processes.

Montreal Protocol

The Montreal Protocol on Substances that Deplete the Ozone Layer is a globally coordinated regulatory action that seeks to phase out nearly 100 ozone-depleting substances (ODS). All 197 UN Member States have ratified this multilateral environmental agreement, the only international treaty to have universal membership, and both developed and developing countries are working to achieve the Protocol's objectives. It entered into force in January 1989. ODSs include hydrofluorocarbons (HCFCs), chlorofluorocarbons (CFCs), halons, methyl bromide, carbon tetrachloride and methyl chloroform and hydrofluorocarbons (HFCs). All are potent greenhouse gases that contribute to climate change (https://www.unenvironment.org/ozonaction/who-we-are/about-montrealproto col).

International Agency for Research on Cancer (IARC)

IARC is a component of the World Health Organization (WHO), located in Lyon, France. The goal of IARC is to promote trans-national interdisciplinary research on cancer with an emphasis on understanding factors that cause cancer burden worldwide and devising detection and prevention strategies. The IARC Monographs Program is a core element of the Agency's portfolio of activities, with international expert working groups evaluating the evidence of the carcinogenicity of specific exposures. Resources also include the publications *Cancer Incidence in Five Continents* series and *GLOBOCAN*.

Organization for Economic Cooperation and Development (OECD)

The Organization for Economic Co-operation and Development (OECD) came into force in 1961, when the United States and Canada joined European members organized under a United States Marshall Plan for reconstruction of Europe after World War II. Thirty-nine member nations and special partners cooperate across the globe, representing over 80% of world trade and investment. Its mission is to promote economic growth, prosperity, and sustainable development, and reducing chemical pollution has been part of its focus on resource efficiency and transition to a circular economy. Safety, Health, and the Environment are components in the OECD framework for measuring well-being, and the organization produces reports on global chemical challenges such as perfluorinated compounds, and hosts conferences like the 2018 Global Forum on Environment, where plastic pollution and the circular economy were themes. The organization also maintains a database on resource productivity and the transboundary movement of wastes, and produces policy documents and white papers on global material resources and state of the economy. An example is the *Global Material Resouces Outlook to 2060: Economic Drivers and Environmental Consequences*, 2018.

The Globally Harmonized System (GHS) of Classification and Labeling of Chemicals is one of the more important international protocols considered in these studies. It was adopted by the UN Economic Commission for Europe (UNECE) in 2002, and comprises standards and specifications for classifying, assessing and labeling chemicals in a universal format during handling, transport and use. It was intended to facilitate more efficient trade, but because of its basis in chemical hazard, rather than risk, has formed the foundation for a variety of human and environmental health assessment tools, like the proprietary GreenScreen platform developed by Clean Production Action (CPA).

The GHS contains a standard specification for a Safety Data Sheet (SDS). The SDS follows a 16 section format which identifies substances and formulas, physical and toxicological hazards to human and environmental health, and outlines safety protocols for accidents and routine handling as well as pertinent regulations associated with the chemical. While used widely, the GHS has still not been implemented by 120 countries, however.

The European Union manages chemicals primarily through the European Chemical Agency (ECHA) under the authority of Registration, Evaluation and Authorization and Restriction of Chemicals (REACH), the Classification, Labeling and Packaging (CLP) and Biocidal Products (BPR) regulations. In contrast to the United States, these regulations place a greater burden on companies to prove the avoidance of harm in products before they are brought to market. Specific regulations exist for specific families of products such as Fertilisers, Detergents, Explosives, Pyrotechnic Articles, Drug Precursors. To comply with REACH regulations companies must identify and manage the risks linked to the substances they manufacture and market, demonstrate how the substance can be safely used, and communicate risk management measures to the users. REACH regulations establish collection and assessment procedures, registration and sector cooperation, and evaluations for compliance. Member state authorities and ECHA assess the manageability of risks associated with specific substances and authorities can restrict a substance's use or ban it altogether.

CLP regulations require companies to classify, label and package substances and mixtures per ECHA prior to placing on the market. The regulations pertain to making judgments about the hazard of substances and mixtures and complying with proper labeling and safety precautions, such as universal pictograms and risk and safety statements that align with GHS. It came into force in 2009. From 2020 onward a 16-digit Unique Formula Identifier (UFI) appears on labels.

In addition to regulation, ECHA provides resources and support for safer substitutions of chemicals of concern. Guidance on alternatives assessment, supply chain intervention, and funding and technical support are provided.

The EU chemicals legislation finder, EUCLEF, enables one to find out how substances are regulated in the EU and the legal obligations of the regulations. This March 2020, instituted service from ECHA lists 40 pieces of EU chemicals legislation.

United States

In the United States, there is no comprehensive national policy for the regulation and management of chemicals. Such a policy would conceivably comprise a vision and goal to protect human and environmental health, have a comprehensive set of objectives and supporting design and management principles, identify quantifiable targets and serve as the aegis for establishing specific laws that would ensure the compliance needed to achieve those objectives.

Current federal and state chemical regulations within the United States can perhaps be best described as a patchwork of laws with significant overlap of jurisdictions, gaps in protection of citizenry, and public confusion and uncertainty. These laws are based on specific known threats to health or standards for products or activities.

The legislation that has most directly addressed chemicals of concern throughout their life cycle within the supply chain is the Toxic Substances Control Act (TSCA), officially named The Frank R. Lautenberg Chemical Safety for the 21st Century Act. The original 1976 law was updated and finally signed into law in 2016. TSCA is administered by the US Environmental Protection Agency (EPA) and regulates chemicals used commercially in the United States. TSCA does not regulate pesticides, chemicals used in cosmetics and personal care products, food, food packaging, or pharmaceuticals.

The US EPA also regulates pesticide levels in food and hazardous waste under the authority of the Federal Insecticide, Fungicide and Rodenticide Act (FIFRA), 1910, revised 1972, and the Resource Conservation and Recovery Act, 1976. The US Food and Drug Administration (FDA) oversees a wide variety of substances under the authority of the Federal Food Drug and Cosmetic Act (FFDCA). The Consumer Products and Safety Commission (CPSC) has some authority over chemicals in products based on the Consumer Product Safety Act, 1972, the Consumer Product Safety Improvement Act (CPSIA), 2008, and the Federal Hazardous Substances Act, 1960, as well as various other laws associated with individual products like drywall and flammable fabrics. As can be see below, there is significant shared authority within the acts, based on technical expertise and the overall mandates of the regulatory agencies.

US Environmental Protection Agency (EPA)

The US EPA is the main administrative organization overseeing compliance with laws related to chemicals in the United States. It was established in 1970, following the enactment of the National Environmental Policy Act (NEPA) and is organized in several major divisions and administered in 10 regions of the country. The Office of Chemical Safety and Pollution Prevention (OCSPP) administers the following laws:

Federal Insecticide, Fungicide, and Rodenticide Act (FIFRA) (https://www.epa.gov/laws-regulations/summary-federal-insecticidefungicide-and-rodenticide-act).

Federal Food, Drug and Cosmetic Act (FFDCA) (https://www.epa.gov/laws-regulations/summary-federal-food-drug-andcosmetic-act).

Toxic Substances Control Act (TSCA) (https://www.epa.gov/laws-regulations/summary-pollution-preventionact).

Pollution Prevention Act, and (https://www.epa.gov/laws-regulations/summary-pollution-prevention-act)

portions of other statutes, like the Endangered Species Act (ESA), and the Food Quality Protection Act (FQPA).

The Office of Chemical Safety and Pollution Prevention (OCSPP)

OCSPP is further organized into the following three sections:

Office of Pesticide Programs (OPP) OPP regulates the manufacture and use of all pesticides (including insecticides, herbicides, rodenticides, disinfectants, sanitizers, and more) in the United States and establishes maximum levels for pesticide residues in food. It also manages the Pesticide Environmental Stewardship Program, a voluntary private and public partnership dedicated to reducing pesticide use and risk, and the Integrated Pest Management in Schools program.

The OPP administers the following laws:

the Federal Insecticide, Fungicide, and Rodenticide Act (FIFRA) (https://www.epa.gov/laws-regulations/summary-federal-insecticide-fungicide-and-rodenticide-act).

the Pesticide Registration Improvement Extension Act (PRIA 3) (https://www.epa.gov/pria-fees).

key parts of:

Food Quality Protection Act (FQPA) (https://www.epa.gov/laws-regulations/summary-food-quality-protectionact).

Federal Food, Drug, and Cosmetic Act (FFDCA) (https://www.epa.gov/laws-regulations/summary-federal-food-drug-andcosmetic-act).

Endangered Species Act (ESA) (https://www.epa.gov/laws-regulations/summary-endangered-species-act).

Office of Pollution Prevention and Toxics (OPPT) OPPT manages programs under the Toxic Substances Control Act and the Pollution Prevention Act (TSCA). Under these laws, EPA evaluates new and existing chemicals and their risks, and finds ways to prevent or reduce pollution before it gets into the environment. It also manages a variety of environmental stewardship programs that encourage companies to reduce and prevent pollution. OPPT implements the Toxic Substances Control Act (TSCA), the Pollution Prevention Act and Section 313 of the Emergency Planning and Community Right-to-Know Act (EPCRA) (https://www.epa.gov/laws-regulations/summary-toxic-substances-controlact; https://www.epa.gov/lawsregulations/summary-pollution-prevention-act; https://www.epa.gov/epcra/what-epcra).

This office manages 18 different programs and projects, including the Toxics Release Inventory (TRI), Safer Choice, Green Chemistry and

Engineering, the TSCA Inventory, new chemicals, import/export and biotech; Pollution Prevention (P2), and Chemical Data Reporting.

Office of Science Coordination and Policy (OSCP) OSCP provides coordination, leadership, peer review, and synthesis of science and science policy within OCSPP and runs the following programs:

Endocrine Disruptor Screening Program (https://www.epa.gov/endocrine-disruption; https://www.epa.gov/endocrine-disruption).

Scientific Advisory Panel (SAP) (https://www.epa.gov/sap).

TSCA Peer Review Committees.

Chemical Safety Advisory Committee (CSAC) (https://www.epa.gov/tsca-peer-review).

Science Advisory Committee on Chemicals (SACC).

The US EPA provides a range of resources and programs that more proactively promote the use of safer chemicals. Chief among them is the Safer Choice program which had replaced the Design for the Environment (DfE) program in 2015. This program was conceived and established as a collaborative, nonregulatory initiative to help companies design and produce commercial products and processes while considering human health and environmental, as well as technological and economic, performance parameters. This program sponsors resource development like alternative assessment and life cycle assessment guides, a standard for safer chemicals (Safer Choice Standard), a certification program, based on the standard, with its own label for household and commercial products, and an annual award program for partners. Consumers can search for safer products and individual chemicals on the program's website.

The Safer Chemical Ingredient List (SCIL) is arranged by functional category and ranks safer chemicals with four ratings from low concern to not yet meeting the standards for the Safer Choice Label. It is based on the Criteria for Safer Chemical Ingredients set by the US EPA. Both the Safer Choice Standard and the Criteria for Safer Chemical Ingredients address a wide range of toxicological concerns: carcinogens, mutagens, reproductive or development toxicants, persistent, bioaccumulative and toxic (PBT) compounds, asthmagens, sensitizers, and endocrine disrupters, as well as a limit on general impurities allowed in a formula. The US EPA also provides several databases for citizenry, including Chemview, a centralized database of information on chemicals subject to TSCA, including EPA assessments, regulatory actions, and health and safety data, and CleanGredients, a list of chemicals, arranged by component class, that meet the Safer Choice Criteria.

The US EPA also manages a pollution prevention program and maintains the Toxics Release Inventory (TRI).

The Pollution Prevention (P2) program provides funding and technical guidance for collaborative projects that pursue source reduction in industry and the commercial sector. Pollution prevention is defined by the EPA as *"reducing or eliminating waste at the source by modifying production processes, promoting the use of nontoxic or less toxic substances, implementing conservation techniques, and reusing materials rather than putting them into the waste stream."*

The Pollution Prevention Act, upon which the P2 program is based, defines "source reduction" to mean any practice which:

Reduces the amount of any hazardous substance, pollutant, or contaminant entering any waste stream or otherwise released into the environment (including fugitive emissions); prior to recycling, treatment or disposal; and

Reduces the hazards to public health and the environment associated with the release of such substances, pollutants, or contaminants.

Reduction can include modification to equipment or technology, process or procedure, the reformulation or redesign of products; the substitution of raw materials and improvements in housekeeping, maintenance, training, or inventory control.

The TRI is a resource for learning about toxic chemical releases and pollution prevention activities reported by industrial and federal facilities since 1987. The purpose of this database is to support informed decision-making by communities, government agencies, companies, and individual citizens. It is mandated by the 1986 Emergency Planning and Community Right-to-Know Act (EPCRA) (https://www.epa.gov/epcra).

There are currently 755 individually listed chemicals and 33 chemical categories covered by the TRI Program. In general these listed chemicals include those in the category of carcinogenic or chronic human health threats, or significant and adverse effects on human or environmental health. Facilities that manufacture, process or otherwise use these chemicals in amounts above established levels must submit annual reporting forms for each chemical. Note that the TRI chemical list doesn't include all toxic chemicals used in the United States.

For each facility there is a link to summarized TRI information for years reported, Federal Registry System (FRS) facility information, a corresponding Risk Screening Environmental Indicator (RSEI) report that provides a quantitative, relative estimate of risk posed by the facility based on the chemical released and potential exposure pathways, and a Pollution Prevention (P2) report presenting measures taken to prevent pollution and reduce the amount of toxic chemicals entering the environment.

Several other U.S. government entities should be mentioned here:

The Occupational Safety and Health Administration (OSHA)

OSHA is an agency of the United States Department of Labor, established under the Occupational Safety and Health Act, signed into law in 1970. Its mission is to ensure safe and healthful working conditions for working men and women by setting and enforcing standards and by providing training, outreach, education and assistance. OSHA inspects and enforces workplace safety measures including those associated with hazardous chemicals and indoor air quality. This involves requirements and review of Safety Data Sheets (SDSs) and adherence to the Hazard Communication Standard (HCS). Under the HCS, the label preparer must provide the identity of the chemical and the appropriate hazard warnings on the SDS. This includes using the eight health threat pictograms of the GHS.

The National Institute for Occupational Safety and Health (NIOSH)

NIOSH was also established in 1970, under the Occupational Safety and Health Act, 1970, and focuses on research in and promotion of worker safety and health as part of the U.S. Centers for Disease Control and Prevention (CDC), in the Department of Health and Human Services. It is comprised of experts in epidemiology, medicine, nursing, industrial hygiene, safety, psychology, chemistry, statistics, economics, and many branches of engineering. The Cancer, Reproductive, Cardiovascular and Other Chronic Disease Prevention Program is an example of work directly related to chemicals and public health.

The National Institute of Environmental Health Sciences (NIEHS)

The National Institute of Environmental Health Sciences conducts research into the effects of the environment on human disease, as one of the 27 institutes and centers of the National Institutes of Health of the Department of Health and Human Services. The mission of the institute is to understand the burden of human illness and disability as caused by the environment. This has entailed research on the dangers of asbestos and lead, for example.

State of California

The State of California has adopted green chemistry as a specific strategy for protecting the health of humans and the environment, and pursues this strategy through two sections within its Environmental Protection Agency, the Department of Toxic Substances Control (DTSC) and its Office of Environmental Health Hazard Assessment (OEHHA).

In 2008, the California legislature enacted the Green Chemistry Initiative by passing two pioneering laws designed to protect citizens from toxic chemicals in products and to provide the public with more information about chemical hazards, and shift more of the burden of proof of safety to industry.

Assembly Bill 1879 created the Safer Consumer Products Program, requiring DTSC to evaluate chemicals of concern in products and their potential alternatives, and to reduce the hazards of chemicals in products.

Senate Bill 509 established a Toxics Information Clearinghouse (TIC) for data on chemical hazards to be used by DTSC in its Safer Consumer Products Program. The Office of Environmental Health Hazard Assessment (OEHHA) was required to identify the hazard traits to be included in the TIC. These were issued in 2012, and specified "... *the hazard traits, toxicological and environmental endpoints and other relevant data to be included...*" in the TIC. The DTSC uses the TIC to identify chemicals of concern as part of its administration of Green Chemistry Initiative and Safer Consumer Products Alternatives regulation. The TIC includes references to five categories of information: chemical, exposure, public health/toxicology, environmental, and regulatory/green chemistry/alternatives.

DTSC promotes safer products by following a four-step regulatory process.

The department compiles a list of Candidate Chemicals of concern; targets Priority Products containing some of the most impactful of these chemicals; guides manufacturers in performing alternatives assessment for these Priority Products and the chemicals of concern within them; and designs any necessary regulations or enforcement to ensure the protection of public and environmental health and safety. Manufacturers must notify DTSC when their products have been identified as Priority Products and work, in the alternatives assessment, to limit the exposure or level of chemicals of concern in their products. If they are unsuccessful or incompliant, DTSC may seek regulatory remedies.

The Green Chemistry Initiative laws were built upon a foundation of other chemical laws, many of them enacted in response to the lack of comprehensive federal regulation, and the desire to protect Californians' health and environment. They include: the Safe Drinking Water and Toxic Enforcement Act (1986), also known as Prop 65; the Children's Environmental Health Protection Act (1999); the California Environmental Contaminant Biomonitoring Program (2006); and other laws directly addressing toxic air contaminants and specific pollutants.

Chemicals management in the marketplace

The private, for profit and not-for profit sectors also engage in extensive chemical management, engaging in research and development, self-regulation, precompetitive cooperation, and education and advocacy. Additionally these entities engage in cooperation/coordination with government policies and regulations. The following are a few examples of industry associations that focus on Green Chemistry:

Zero Discharge of Hazardous Chemicals (ZDHC)

ZDHC is an international consortium of member organizations from the textile, apparel, leather and footwear industries established in 2011. These include brands, value chain affiliates and associates. All have committed to working collaboratively toward zero discharge of hazardous chemicals from their manufacturing processes. The nonprofit group focuses on value chain intervention within its member industries and encourages the participation of formulators, suppliers and brands. It produces chemical guidance and best practice sheets, maintains a Manufacturing Restricted Substances List (MRSL), and develops audit protocols, trainings, and capacity-building tools. It has established MRSL conformance standards for companies to meet, leveraging existing third-party certification services. The group is guided in strategy and action by the Roadmap to Zero Program.

Green Chemistry and Commerce Council (GC3)

GC3 is a U.S. multi-stakeholder collaborative with a mission to drive the commercial adoption of green chemistry by initiating and guiding action across industries, sectors and supply chains. The Green Chemistry & Commerce Council (GC3) was established in 2005, and is a project of the Lowell Center for Sustainable Production at the University of Massachusetts, Lowell. It comprises individual companies and academic and nonprofit members, holds an annual Innovators' Roundtable and coordinates several working groups, in retail, education, mainstreaming, and innovation. The group produces resource materials, holds trainings, and supports a Startup Network of innovative companies developing greener materials (https://www.uml.edu/Research/Lowell-Center/default.aspx).

Interstate Chemicals Clearinghouse (IC2)

The Interstate Chemicals Clearinghouse (IC2) is an association of state, local, and tribal governments in the United States that promotes a clean environment, healthy communities, and a vital economy through the development and use of safer chemicals and products. The IC2 is a program of the Northeast Waste Management Officials' Association (NEWMOA), which provides management and staff support for IC2 and serves as its fiscal agent (http://www.newmoa.org/).

NEWMOA is a nonprofit, nonpartisan, interstate association whose membership is composed of the state environment agency programs that address pollution prevention, toxics use reduction, sustainability, materials management, hazardous waste, solid waste, emergency response, waste site cleanup, underground storage tanks, and related environmental challenges in the northeast states.

Membership in IC2 is national and also comprises nonprofit, academic, and industry supporting members. The group coordinates regulatory efforts, capacity building, and the sharing of data and information. It has produced an Alternatives Assessment Guide and shares intelligence on alternatives assessment, chemicals of concern, and hazard assessment.

American Chemical Society Green Chemistry Institute (ACS /GCI)

The mission of the ACS GCI is to catalyze and enable the implementation of green and sustainable chemistry and engineering throughout the global chemical enterprise and across the Society. It was established in 1997 as an independent nonprofit and joined the ACS in 2001. Its strategic mission comprises three areas of endeavor: science, education, and industry and the organization works to promote discovery and innovation for sustainability, develop education and communication of the principles of green chemistry, and foster the acceleration of adoption of green chemistry in all sectors.

Green America, Clean Electronics Production Network (CEPN)

CEPN is a multi-stakeholder innovation network made up of suppliers, brands, labor and environmental advocates and other experts working toward reducing and eliminating toxic chemical exposure to workers in the production of electronics. Formed by the Center for Sustainability Solutions at Green America in 2016, the CEPN focuses on five substantive areas: Worker Empowerment and Engagement, Tracking and Monitoring Exposures, Qualitative Exposure Assessment, Targeted Safer Substitutions, and Standardized Process Chemicals Data Collection.

Clean Production Action (CPA)

CPA is an organization that runs five different programs in service to a mission to design and deliver strategic solutions for green chemicals, sustainable materials, and environmentally preferable products. The BizNGO Working Group for Safer Chemicals and Sustainable Materials (BizNGO) is a collaboration of business and environmental leaders working together to define and implement innovation in and adoption of safer chemicals and sustainable materials, driving a market transition to a healthier economy. The CFP aims to improve global chemical use by measuring and disclosing data on business progress toward safer chemicals. It provides an internal assessment survey for companies to benchmark their efforts in reducing their chemical footprint. GreenScreen for Safer Chemicals is a GHS-based tool designed to assess and benchmark chemicals based on hazard. The Investor Environmental Health Network (IEHN) is a membership-based, investor collaborative that promotes the use of safer chemicals to enhance shareholder value, public health, and the environment.

The International Pollutants Elimination Network (IPEN)

IPEN is an international network of public interest NGOs working to eliminate the most hazardous chemicals from production and use, educate the public about best practices of chemicals management, raise the regulatory standards for chemicals, halting the spread of toxic metals and building a toxics-free advocacy network.

Summary

Green Chemistry, the practice of eliminating or reducing hazardous chemicals from products and processes, is an essential part of global efforts toward more sustainable design, manufacturing, and use. The field exists within a broader context of concerns about the current climate crisis, growing scarcity of raw materials, and threats to human and environmental health. These are pressing and existential threats to life on earth.

Efforts to address these universal concerns are driven by participants from government, private enterprise and the not-for-profit sector, each with different, and sometimes competing, objectives. They have taken the form of policies, regulations, cooperative agreements, and financial incentives for discovery and innovation at the international, national and local levels. All of these approaches present their own opportunities and constraints and combine in complex ways to create a cultural context that any green chemistry project manager must acknowledge and work to her advantage.

In the following chapters, we lay out a process for assessment, analysis and substitution of hazardous chemicals with safer choices and four case studies across the industry sectors of household products, apparel, additive manufacturing, and energy. In each of our examples, different technical and EHS factors might have come to the fore, but the broad societal challenges outlined in this chapter, form a consistent and compelling context.

References

Burek, P., Satoh, Y., Fischer, G., Kahil, M., Scherzer, A., Tramberend, S., Nava, L. F., Wada, Y., Eisner, S., Flörke, M., Hanasaki, N., Magnuszewski, P., Cosgrove, B., & Wiberg, D. (2016). Water futures and solution - fast track initiative (final report). *Environmental Science.*

Erythropel, H. C., Zimmerman, J. B., de Winter, T. M., Petitjean, L., Melnikov, F., Lam, C. H., ... Anastas, P. T. (2018). The green ChemisTREE: 20 years after taking root with the 12 principles. *Green Chemistry, 20*(9), 1929−1961.

Richey, A.S., Thomas, B.F., Lo, M.-H., Reager, J.T., Famiglietti, J.S., Voss, K.,. & Rodell, M. (2015). Quantifying renewable groundwater stress with GRACE. *Water Resource Re*search, *51*(7), 5217−5238. https://doi.org/10.1002/2015WR017349.

Further reading

Alexandratos, N., & Bruinsma, J. (2012). *World agriculture towards 2030/2050: The 2012 revision, ESA working paper No. 12−03.*

Anastas, P. T., & Warner, J. C. (1998). *Green chemistry: Theory and practice.* New York: Oxford University Press.

UNEP, Arendal, G. (2016). *Marine litter vital graphics. United Nations Environment Programme and GRID-Arendal. Nairobi and Arendal.*

aus der Beek, T., Weber, F. A., Bergmann, A., Gruttner, G., & Carius, A. (2016). *Pharmaceuticals in the environment: Global occurrence and potential cooperative action under the strategic approach to international chemicals management. 94.* Germany Federal Environmental Agency.

Benbrook, C. M. (2016). Trends in glyphosate herbicide use in the United States and globally. *Environmental Sciences Europe, 28*(1), 1—15.

Crossman, J., Hurley, R. R., Futter, M., & Nizzetto, L. (2020). Transfer and transport of microplastics from biosolids to agricultural soils and the wider environment. *Science of the Total Environment, 724,* 138334.

Esdaile, L. J., & Chalker, J. M. (2018). The mercury problem in artisanal and small-scale gold mining. *Chemistry—A European Journal, 24*(27), 6905—6916.

T. Fennelly & Associates. (2015). *Advancing Green Chemistry: Barriers to adoption & ways to accelerate green chemistry in supply chains.* Available at http://greenchemistryandcommerce. org/assets/media/images/Publications/Advancing-Green-Chemistry-Report-15-June-2015.pdf.

Fernandez-Cornejo, J., Nehring, R. F., Osteen, C., Wechsler, S., Martin, A., & Vialou, A. (2014). *Pesticide use in US agriculture: 21 selected crops, 1960—2008. USDA-ERS Economic information Bulletin, (124).*

GBD 2015 Risk Factors Collaborators. (2016). Global, regional, and national comparative risk assessment of 79 behavioural, environmental and occupational, and metabolic risks or clusters of risks, 1990—2015: A systematic analysis for the global burden of disease study 2015. *Lancet (London, England), 388*(10053), 1659.

Global chemicals outlook IIFrom legacies to innovative solutions: Implementing the 2030 Agenda for sustainable development Copyright © United Nations environment Programme.(2019). Available at: https://www.unep.org/explore-topics/chemicals-waste/what-we-do/policy-and-governance/global-chemicals-outlook.

HEI Special Report 20, Burden of Disease Attributable to Coal-Burning and Other Air Pollution Sources in China.(2020). Available at https://www.healtheffects.org/publication/gbd-air-pollution-india.

Hudson-Edwards, K. A., Jamieson, H. E., & Lottermoser, B. G. (2011). Mine wastes: Past, present, future. *Elements, 7*(6), 375—380.

Landrigan, P. J., Fuller, R., Acosta, N. J., Adeyi, O., Arnold, R., Baldé, A. B., ... Zhong, M. (2018). The Lancet Commission on pollution and health. *The Lancet, 391*(10119), 462—512.

McDonald, B. C., DeGouw, J. A., Gilman, J. B., Jathar, S. H., Akherati, A., Cappa, C. D., ... Trainer, M. (2018). Volatile chemical products emerging as largest petrochemical source of urban organic emissions. *Science, 359*(6377), 760—764.

OECD. (2019). *Global material resources outlook to 2060 economic drivers and environmental Consequences. OECD Publishing.*

Outlook, E. (2010). *International Energy Outlook.* Outlook. Available at www.eia.gov/oiaf/ieo/index.html.

Pellizzari, E. D. (1987). *Total Exposure Assessment Methodology (TEAM study) (Vol. 1). US Environmental Protection Agency, Office of Research and Development.*

Richey, A. S., Thomas, B. F., Lo, M. H., Famiglietti, J. S., Swenson, S., & Rodell, M. (2015). Uncertainty in global groundwater storage estimates in a total groundwater stress framework. *Water Resources Research, 51*(7), 5198—5216.

de Souza Machado, A. A., Kloas, W., Zarfl, C., Hempel, S., & Rillig, M. C. (2018). Microplastics as an emerging threat to terrestrial ecosystems. *Global Change Biology, 24*(4), 1405—1416.

de Souza Machado, A. A., Lau, C. W., Till, J., Kloas, W., Lehmann, A., Becker, R., & Rillig, M. C. (2018). Impacts of microplastics on the soil biophysical environment. *Environmental Science & Technology, 52*(17), 9656—9665.

Tickner, J. A., & Becker, M. (2016). Current opinion in green and sustainable chemistry. *Chemistry, 2*, 3.

Toensmeier, E., & Blake, A. (2017). *Industrial perennial crops for a Post-petroleum materials economy.*

Vanham, D., Hoekstra, A. Y., Wada, Y., Bouraoui, F., De Roo, A., Mekonnen, M. M., ... Bidoglio, G. (2018). Physical water scarcity metrics for monitoring progress towards SDG target 6.4: An evaluation of indicator 6.4. 2 "Level of water stress". *Science of the Total Environment, 613*, 218−232.

Wada, Y., Flörke, M., Hanasaki, N., Eisner, S., Fischer, G., Tramberend, S., ... Wiberg, D. (2016). Modeling global water use for the 21st century: The Water Futures and Solutions (WFaS) initiative and its approaches. *Geoscientific Model Development, 9*(1), 175−222.

http://www.basel.int/Home/tabid/2202/Default.aspx.

http://www.centerforsustainabilitysolutions.org/clean-electronics#cepn-about.

http://www.fao.org/home/en/.

http://www.healthdata.org/sites/default/files/files/policy_report/2019/GBD_2017_Booklet.pdf.

http://www.navigantresearch.com/wp-content/uploads/2011/06/GCHEM-11-Executive-Summary.

http://www.pic.int/TheConvention/Overview.

http://www.pops.int.

http://www.pops.int.

http://www.saicm.org/About/Texts/tabid/5460/Default.aspx.

https://archive.epa.gov/epa/tsca-peer-review/chemical-safety-advisory-committee-csac-basic-information.html.

https://bcgc.berkeley.edu.

https://bcorporation.net/welcome-sdg-action-manager.

https://calepa.ca.gov.

https://chemview.epa.gov/chemview/.

https://dtsc.ca.gov.

https://dtsc.ca.gov/scp/.

https://dtsc.ca.gov/scp/toxics-information-clearinghouse/.

https://echa.europa.eu.

https://echa.europa.eu/regulations/biocidal-products-regulation/understanding-bpr.

https://echa.europa.eu/regulations/clp/understanding-clp.

https://echa.europa.eu/regulations/reach/understanding-reach.

https://ipen.org/documents/citizens-report-saicm-implementation-2012-2015.

https://monographs.iarc.who.int/agents-classified-by-the-iarc/.

https://nca2018.globalchange.gov (online report Fourth National Climate Assessment, Volume II Impacts, Risks and Adaption in the United States).

https://oehha.ca.gov.

https://unece.org/DAM/trans/danger/publi/ghs/ghs_rev04/English/ST-SG-AC10-30-Rev4e.pdf.

https://unfoundation.org/what-we-do/issues/sustainable-development-goals/?gclid=Cj0KCQjwo-aCBhC-ARIsAAkNQiuq2tY0gHsp74Ix9dipoZxsvVEzM DhrlcR6HxFTBxbBh5HEl6vsQWEaAvJLEALw_wcB.

https://www.acs.org/content/acs/en/greenchemistry/about.html.

https://www.cdc.gov/niosh/programs.html.

https://www.chemistryworld.com/news/water-purifier-harnesses-green-chemistry/3004277.article.

https://www.cleanproduction.org.

https://www.cleanproduction.org/programs/greenscreen.

https://www.congress.gov/bill/116th-congress/senate-bill/483/text.

https://www.cpsc.gov.

https://www.cpsc.gov/Business-Manufacturing/Business-Education/Business-Guidance/FHSA-Requirements.

https://www.cpsc.gov/Regulations-Laws-Standards/Statutes/Summary-List/Consumer-Product-Safet-Act.

https://www.cpsc.gov/Regulations-Laws-Standards/Statutes/The-Consumer-Product-Safety-Improvement-Act.

https://www.epa.gov/aboutepa/about-office-chemical-safety-and-pollution-prevention-ocspp.

https://www.epa.gov/aboutepa/about-office-chemical-safety-and-pollution-prevention-ocspp.

https://www.epa.gov/assessing-and-managing-chemicals-under-tsca/frank-r-lautenberg-chemical-safety-21st-century-act.

https://www.epa.gov/chemical-data-reporting.

https://www.epa.gov/endocrine-disruption/endocrine-disruptor-screening-program-edsp-overview.

https://www.epa.gov/enforcement/federal-insecticide-fungicide-and-rodenticide-act-fifra-and-federal-facilities.

https://www.epa.gov/epcra.

https://www.epa.gov/greenchemistry.

https://www.epa.gov/indoor-air-quality-iaq/volatile-organic-compounds-impact-indoor-air-quality.

https://www.epa.gov/laws-regulations/summary-federal-insecticide-fungicide-and-rodenticide-act.

https://www.epa.gov/laws-regulations/summary-food-quality-protection-act.

https://www.epa.gov/laws-regulations/summary-food-quality-protection-act.

https://www.epa.gov/laws-regulations/summary-national-environmental-policy-act.

https://www.epa.gov/laws-regulations/summary-pollution-prevention-act.

https://www.epa.gov/laws-regulations/summary-resource-conservation-and-recovery-act.

https://www.epa.gov/p2.

https://www.epa.gov/pesticide-contacts/office-pesticide-programs-contacts-division-and-topic.

https://www.epa.gov/saferchoice/history-safer-choice-and-design-environment.

https://www.epa.gov/saferchoice/safer-ingredients.

https://www.epa.gov/saferchoice/standard.

https://www.epa.gov/sap.

https://www.epa.gov/toxics-release-inventory-tri-program.

https://www.epa.gov/tsca-inventory.

https://www.epa.gov/tsca-peer-review.

https://www.fda.gov.

https://www.fda.gov/regulatory-information/laws-enforced-fda/federal-food-drug-and-cosmetic-act-fdc-act.

https://www.federalregister.gov/documents/2020/08/17/2020-17903/tsca-science-advisory-committee-on-chemicals-request-for-nominations-extension-of-nomination-and.

https://www.fws.gov/endangered/laws-policies/.

https://www.greenchemistryandcommerce.org/about-gc3/what-is-the-gc3.

https://www.healtheffects.org/publication/burden-disease-attributable-coal-burning-and-other-air-pollution-sources-china.

https://www.iarc.who.int.

https://www.ipcc.ch/sr15/.

https://www.linkedin.com/company/zdhc-foundation-zero-discharge-of-hazardous-chemicals.

https://www.niehs.nih.gov.

https://www.oecd-ilibrary.org/economics/oecd-economic-outlook/volume-2020/issue-2_39a88ab1-en.

https://www.oecd.org/environment/waste/.

https://www.osha.gov.

https://www.theic2.org/about_ic2.

https://www.wbcsd.org/Programs/People/Sustainable-Development-Goals/Resources/Chemical-Sector-SDG-Roadmap.

https://www.who.int/health-topics/air-pollution#tab=tab_1.

www.roadmaptozero.com.

Chapter 2

From education to implementation: the Greener Solutions course at the Berkeley Center for Green Chemistry

The Berkeley Center for Green Chemistry

The Berkeley Center for Green Chemistry (BCGC) has been promoting the development and use of safer alternatives to toxic chemicals since its founding in 2012. Through education, research, and engagement it has worked with over two dozen organizations to replace chemicals of concern, educating over 100 students in its Greener Solutions graduate course, and has funded over 30 graduate researchers through its fellowship and internship programs.

The mission of the Berkeley Center for Green Chemistry is to bring about a generational transformation toward the design and use of inherently safer chemicals and materials. The founders believe that embedding the principles of green chemistry into science, markets and public policy will provide the foundation for safeguarding human health and ecosystems and provide a cornerstone for a sustainable, clean energy economy.

The center is located in the College of Chemistry at the University of California at Berkeley. It draws support from students and faculty from five different colleges or schools on campus, including Public Health, Engineering, Natural Resources, Law, and Business. It is administered by an executive director who reports to an academic director in the Department of Chemistry and is guided by a nine-member advisory board comprised of faculty and professionals. It has been supported financially by university research funds, government grants, private sector gifts, and consulting income.

Many of the research centers at UC Berkeley have been formed because of the need to work across disciplinary boundaries in order to solve complex problems. BCGC is among these: the development of safer alternative

Green Chemistry in Practice. https://doi.org/10.1016/B978-0-12-819674-8.00010-2

materials and processes that will be adopted by organizations within our current economic and cultural context requires expertise in a wide range of subjects.

The center was established on the premise that assembling trans-disciplinary teams of researchers and analysts appropriate to each individual challenge is the most successful approach to replacing harmful chemicals with safer alternatives. Moreover, its founders realized that the adoption of safer practices must be pursued within different societal venues and involve three main subject areas: education, research, and engagement, summarized below. Finally, success (adoption of safer materials and processes) will be achieved through a sequence of accomplishments that linked these areas for emergent outcomes. The case studies in this volume are examples of this linkage.

Education

The mission of BCGC requires that its educational activities be pursued in several different venues. Educating the next generation of chemistry, public health, and engineering professionals is primarily and largely executed by formal instruction, but the education of particular industry sectors and the general public is also an important and exciting part of the center's work. Thus, the center employs three delivery methods for education: formal classroom instruction, individual mentorship, and a professional internship program.

The center has hosted a core program of courses that integrate the chemical and environmental health sciences with the study of public and private governance and management. This interdisciplinary program focuses on project and team-driven approaches to solving important material challenges and offers actionable results to industry, government, and nongovernmental organizations. The program has included seven courses and a green chemistry lab program over the span of 8 years. These courses have drawn students from five different colleges within the university. Chief among them have been the School of Public Health, the College of Chemistry, the College of Natural Resources, and the College of Engineering, Likewise, these courses could not have been taught without the collaborative contributions of the faculty from these colleges, as well as experts from the Schools of Law and Business. The Greener Solutions graduate course has been the flagship course within the program, is the main source of the case studies in this volume, and is described in detail below.

Mentorship of individual postgraduate, graduate, and undergraduate students has been an important activity at the center throughout its history and has been supported by dozens of university faculty and staff associated with the center. The center has directly employed several postdoctoral students.

The professional Greener Partnerships internship program was established in 2016. This has been supported by an annual gift from a private donor to run a program for students and recent graduates who are working in areas associated with the center's mission. The program stipulates that they be actively engaged in work being pursued by an off-campus organization and that their funding is matched by that organization, thus encouraging a commitment from all parties toward a tangible, shared objective.

To date the program has sponsored over a dozen students in this program, including a chemist who worked with the software company Autodesk to develop a hazard assessment and health scorecard for stereolithography resins (see Chapter 5 case study), a toxicologist who worked on public health policy issues with the nonprofit Natural Resources Defense Council, two mechanical engineers who worked with the US EPA to develop a comprehensive performance review of additive manufacturing materials and processes (also Chapter 5 case study), and two chemists who worked with the USDA Western Regional Research Center on expanding the applications for reversible bond antimicrobials (see Chapter 3 case study).

In 2020, the center received funding from the US Environmental Protection Agency Pollution Prevention (P2) program, Region 9, to expand this program. This funding supported a 2-year investigation and promotion of alternatives to poly and perfluoroalkyl substances (PFAS) in the carpet and rug and packaging sectors.

The intern program has been important for several reasons: it has extended work started in the Greener Solutions course, expanded BCGC's outreach to industry, government, and nonprofit organizations, and offered novel educational and professional experience opportunities to UC Berkeley students and graduates. Often these experiences have been a pivotal career bridge from academic to professional life.

Research

BCGC funds and guides research in designing novel chemical processes and materials and in investigating new approaches to toxicity testing, exposure analysis, and alternatives assessment. The bulk of this research has been supported by grants from the National Science Foundation (NSF), specifically the Innovations in Graduate Education and Research Training or IGERT program (now succeeded by the NRT program). BCGC has funded 28 graduate fellows for 2 years each, providing tuition and stipends and selective travel funds on a competitive basis. Fellows are accepted based on a rigorous review of research proposals aligned with the mission of the center and recommendations from faculty sponsors from at least two different disciplines. The fellows have come from individual science, engineering, and public health laboratories across campus and have conducted their research on a wide range of subjects, from building a better battery to measuring particulates in indoor air pollution to tracking regulations for bauxite mining in Asia. In addition to supporting work by associated researchers in laboratories across campus, staff and faculty at BCGC conduct their own research associated with outreach and contract funding.

Engagement

BCGC provides technical support to decision-makers, workers, community organizations, and businesses working to advance safety and health through green chemistry. This includes fee-for-services consulting, the hosting of conferences, and volunteer activities in local schools.

BCGC staff and faculty have served on numerous advisory boards of not-for-profit organizations. Two members of the BCGC Advisory Board, Drs. Megan Schwarzman and Ann Blake have served on the California Environmental Protection Agency, Department of Toxic Substances Control Green Ribbon Science Panel. The goals of Article 14 of the California Health and Safety Code, that established the panel are: "... significantly reducing adverse health and environmental impacts of chemicals used in commerce, as well as the overall costs of those impacts to the state's society, by encouraging the redesign of consumer products, manufacturing processes, and approaches." The 15-member board comprises academic and professional experts who advise the department on issues associated with green chemistry such as the implementation of the department's Safer Consumer Products regulations.

Outreach at the Berkeley Center for Green Chemistry has included the convening of symposia and conferences, including a Green Chemistry conference in 2011, a BPA-free Can Lining symposium in 2016, and a SAGE Sunset Celebration symposium in 2018.

Green Chemistry: Collaborative Approaches and New Solutions, 2011, was one in a series of annual conferences funded by the Philomathia Foundation and was BCGC's first national conference. It introduced the collaborative, interdisciplinary approaches piloted by BCGC and featured leaders in these fields, including authors of the seminal work, Principles of Green Chemistry, Dr. Paul Anastas, then assistant administrator for the US EPA's Office of Research and Development and the Science Advisory to the Agency, and Dr. John Warner, president of Warner Babcock Institute for Green Chemistry, and Dr. Robert Grubbs, Nobel Laureate, Chemistry, California Institute of Technology.

The *BPA-free Can Linings* public forum, 2016, was a collaborative effort with the California Office of Environmental Health Hazard Assessment (OEHHA) and Department of Toxic Substances Control (DTSC) of the California Environmental Protection Agency, the US Food, and Drug Administration, and the National Toxicology Division, of the National Institute of Environmental Health Sciences. In this 1 day symposium speakers discussed the challenges and opportunities of bisphenol A as a coating on metal, and food-containing cans, finding functional alternatives and next steps and future trends.

The *SAGE Sunset Celebration*, 2018, was an opportunity to highlight a portion of the research work completed by fellows of the NSF IGERT Systems Approach to Green Energy (SAGE) program over the span of 6 years.

This 1-day symposium was also centered on a theme of multi-disciplinary collaboration and included pairings of collaborators in each of over a dozen presentations, including partners from government, nonprofits, and industry.

The center works with industry to develop safer practices and performance in various sectors through funding from government grants. Currently, the center is working with various government and industry stakeholders and trade associations to proof safer alternatives to poly and perfluoryl alkyl substances (PFASs).

BCGC engages in fee-for-services consulting to organizations when appropriate to its academic mission and independence. The center has consulted with several organizations including Costco Wholesale, Levi Strauss and Company, Autodesk, and the Oakland EcoBlock. For Costco Wholesale, a team of collaborators reviewed the company's chemicals management system for global nonfood items in furniture, textiles, and household products and offered recommendations in the areas of information management, personnel organization, and model chemical control practices, education, and public outreach in 2018.

The center has collaborated and engaged with several green chemistry professional organizations and academic groups, including the Green Chemistry and Commerce Council (GC3), the American Association of Alternatives Assessment (A4), the Ellen MacArthur Foundation (EMF) Disruptive Innovation Festival, and Circular Economy 100 (CE100) group, the Green Chemistry Center of Excellence, York University and the Global Green Chemistry Centers (G2C2) network, and the American Chemical Society (ACS) Green Chemistry Institute. BCGC staff and faculty have been regular speakers at the annual ACS Green Chemistry and Engineering conference.

The Greener Solutions course

Greener Solutions PH271H, School of Public Health, University of California at Berkeley, Industrial Hygiene masters program elective.

Greener Solutions (Public Health 271H) is a 3-unit project-based graduate course that pairs interdisciplinary teams of graduate students and advanced undergraduates from chemistry, engineering, public health, and business with a partner company to research safer alternatives to hazardous chemicals used in a product, material, or manufacturing process.

An interdisciplinary teaching team of three instructors (supplemented by guest lecturers) provides expertise in the fields of chemistry, environmental health, toxicology, bio-inspired design, and chemicals policy. Students consult with formulating chemists and material scientists at their partner companies for technical input on their particular application. Lectures, discussions, and assignments further support the technical aspects of the project, while additional class sessions teach process-oriented skills such as interdisciplinary research methods, strategies for high-performing teams, and effective written and oral communication.

At the end of the semester, students summarize their findings and recommendations in a written report, (typically called an "Opportunity Map"), poster presentation, and formal on-campus presentation to the partner company and the public.

Partner organizations are expected to make available one technically knowledgeable contact person throughout the project to present the organization's background and the initial challenge to the class at its start and attend the presentation at the end. Typically, partners desire more interaction and regular weekly phone conferences are not uncommon. BCGC is part of a public academic institution and has obligations to transparency, publication of results, and intellectual integrity. For these reasons, the course must operate in the precompetitive space, meaning that no proprietary information can be shared by the partner. This means that intellectual property issues and the need for nondisclosure agreements can be avoided.

Students apply to take the course and their CVs are reviewed for capabilities so each successful candidate is placed on a team according to interest and capabilities. Each team typically has complementary skills in chemistry, public health, and engineering at a minimum. Research is done through peer-reviewed journals, interviews, and curatorial lists; no bench-top science experimentation is done, but opportunities for this are available for interested individuals in the Greener Partnerships program.

Course history

The Greener Solutions course has been held for ten fall sessions with 24 partners from 2012 through 2021. It was originally conceived in 2011, and borrowed substantially in its experiential and interdisciplinary approach from several existing courses at UC Berkeley. Challenges have been from a wide range of industry sectors: electronics, apparel, personal care, construction, additive manufacturing, and furniture. Table 2.1 below summarizes the challenges and partners.

Pedagogical principles

Complex problems require interdisciplinary solutions and lesson plans and teams are designed for complementary, multi-disciplinary expertise. The following are some of the guiding pedagogical principles of the course:

"Nine-Way Learning": the three groups involved, faculty, students, and partners, enjoy learning from representatives from the other groups and from members of their own. Interdisciplinary teams are designed to encourage a chemist to instruct an engineer, an industry partner to teach a faculty member, and partner members in an organization to learn more from each other as a result of the Greener Solutions experience.

TABLE 2.1 Greener Solutions project challenges and partner organizations, 2012–21.

Greener Solutions Partners & Projects			
YEAR	**PARTNER ORG.**	**SECTOR**	**CHALLENGE**
2012	Hewlett Packard	Electronics	Identify emerging contaminants from e-waste
2013	Levi Strauss & Co.	Apparel	Alternatives to formaldehyde in permanent press garments
2014	Seventh Generation Method Beautycounter	Personal Care	Safer preservatives for personal care products
	General Coatings CalEPA	Building Materials	Alternatives to spray polyurethane foam insulation
2015	Autodesk	Additive Mfg.	Inherently safer 3D printing resins
	Method Amyris	Household Cleaning	Safer surfactants for low-temperature cleaning products
2016	Steelcase	Furniture Mfg.	Alternative colorant methods for polymer furniture
	Patagonia	Apparel	Inherently safer mosquito-repellant clothing
	Mango Materials	General Mfg.	Biodegradable pigments for marine coatings
2017	W. L. Gore	Fabric Mfg.	Durable water-repellent coating for outerwear
	MycoWorks	Fabric Mfg.	Bio-based treatments for mycelium "leather"
2018	Method	Personal Care	Safer UV blockers for sunscreens
	Method	Household Products	Redesigned packaging to prevent ocean contamination
	Oakland Ecoblock	Building Materials	Safer UV blockers for roofing materials
2019	Nike	Apparel & Shoes	Alternatives to DMF in synthetic leather manufacturing
	US EPA, Region 9	Additive Mfg.	Safer cross-linkers in Stereolithography printing resin
2020	CA Dept. of Toxic Substances Control	Recycling	Removal of PFASs in carpets and rugs at recycling stage
	Method	Food Packaging	Alternatives to PFASs in food packaging
2021	American Assn. of Tire Manufacturers Washington State Dept. of Ecology	Automotive	Alternatives to 6PPD-quinone in tires
	Defend our Health	Household, Commercial Care	Alternatives to PFASs in floor polishes
	Noble Oceans Farms	Food Packaging	Compostable packaging for frozen kelp
	USDA	Food Packaging	Compostable adhesives for PLU stickers

Source: Thomas A. McKeag, 2021 (original material by author).

Project-based Learning: the entire course is focused on a specific real-world objective, so that all forms of instruction, lectures, demonstrations, field trips, and group experiential activities, contribute to this objective and are seen as relevant and worthwhile.

Practicing Professional Practice: the only way to train professionals is to have them practice being professional, and this is reflected in the rigor of content review and in the conductance of the project by the teams.

Constraints Prompt Innovation: this is a course about creative problem-solving and tangible constraints typically force innovative solutions from a wider search for answers.

Balancing Technical and EHS (Environmental Health and Safety) Performance: Both are given comparable consideration and assessed and designed from the start.

Hazard-based rather than risk-based chemical assessment: the course focuses primarily on inherent chemical hazards associated with the material or process, with exposure and population vulnerability as secondary and contextual.

Working at the top of the intervention pyramid: avoidance and replacement of toxic chemicals as far upstream in the value chain as possible is the priority approach with management and engineering controls as secondary choices (Fig. 2.1).

Actionable Recommendations: all recommendations are framed within the context of the partner's needs and capabilities and identified on a spectrum of feasibility from "lead pipe cinch to blue sky pipedream". Moreover, students are encouraged to characterize the capability gaps currently preventing full adoption of a proposed option and detail the steps necessary to fill those gaps.

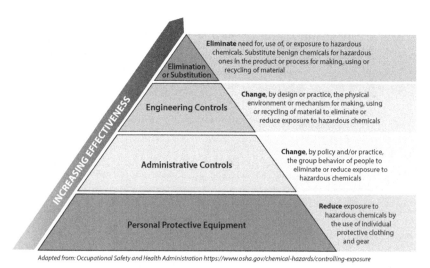

Adapted from: Occupational Safety and Health Administration https://www.osha.gov/chemical-hazards/controlling-exposure

FIGURE 2.1 Chemical hazard intervention pyramid. *Adapted from Occupational Safety and Health Administration. http://chemical-hazards/contolling-exposure*

Studying Innovation: The following three concepts form a nested framework for introducing and inculcating creativity and innovative thinking within the project teams:

Systems Thinking: whole system thinking is encouraged during the entire challenge, particularly during the problem definition and alternatives assessment phases. This is important in avoiding regrettable substitutions that might solve one specific problem while creating others within a manufacturing process. Students are guided by principles from the work of Dr. Donella Meadows and encouraged to consider the overall system outcome, its components, and the relationships between components within the system. From this initial assessment, they proceed to map and characterize interventions in the system they have outlined.

Design Thinking: Design thinking, as popularized by the design firm Ideo, and the Scientific Method have a lot in common and both follow a rational sequence of inquiry and iterative testing of alternatives. In design thinking, STEM students who typically investigate the existing are encouraged to also imagine the possible, engage in divergent and unconstrained brainstorming, and then test alternatives with the criteria for success established in the problem definition phase of the project (Fig. 2.2).

Bio-inspired Design (BID) Innovation: within the design thinking training, biologically inspired design is introduced. BID is the study and translation to technology of biological forms, processes, and systems in order to solve human challenges. Student teams are guided in the translation methods, helped with unfamiliar research pathways, and encouraged to speak with experts in biology. BID-based solutions are not required but encouraged and most teams find inspiration and novel solutions from the rich source field of biology and evolution. BID can be especially useful during the divergent, brainstorming early phase of the project, and systems thinking is reinforced by the referral to the Biodesign Cube, shown in Fig. 2.3. Students are encouraged to form their own customized versions of this template, for example, by adding "time", "material" and "space" to the key parameters list, or more precisely defining the areas of endeavor. Relationships between and among the axes are stressed as part of the systems thinking approach.

Typical course sequence

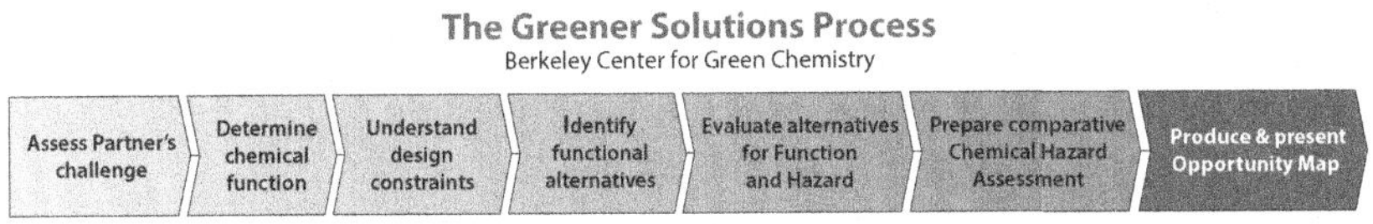

The Greener Solutions Process
Berkeley Center for Green Chemistry

Assess Partner's challenge · Determine chemical function · Understand design constraints · Identify functional alternatives · Evaluate alternatives for Function and Hazard · Prepare comparative Chemical Hazard Assessment · Produce & present Opportunity Map

FIGURE 2.2 The Greener Solutions investigation process. *Source: Thomas A. McKeag, 2021 (original material by author).*

The Bio-design Cube

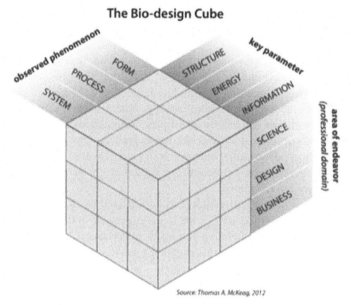

Source: Thomas A. McKeag, 2012

FIGURE 2.3 Bio-inspired design cube. *Source: Thomas A. McKeag, 2008 (original material by author).*

Training Methods: the following is a typical sequence of training found in the course:

1. Problem Definition and Background
   Challenge partner introduction and a formal interview
   Functional performance definition
   Baseline performance metrics
   Opportunities and Constraints
   Contextual Drivers, issues, and concerns
   The content ramp-up in chemistry, public health, and toxicology
   Skills ramp-up in literature search, hazard assessment, project management, communication
2. Development of Strategy Options
   Systems analysis
   Biological inspiration
   Preliminary alternatives assessment
3. Alternatives Assessment
   Final technical and EHS performance metrics
   Comparative chemical hazard assessment
   Ranking of strategy candidates
   Preliminary recommendations and synthesis

4. Communication of Findings
Preparation of Opportunity Map
Inhouse presentation to peers
Public presentation
Issue of Final report
Extensions: papers, conferences, internships

A typical course sequence will start well before actual class time when the instructors will meet with the organizational partner to define the challenge and expected outcomes. This might be the avoidance, replacement, or mitigation of a toxic solvent, a different way to manufacture a product using more benign materials, or a more efficient way to chemically process materials. In all cases, a specific "bad actor" chemical is identified to use as a baseline against which to compare alternatives. At this time, students are submitting their CVs, so that the teaching team can admit them to the course, and more importantly, place them on interdisciplinary teams based on their complementary skills and preferences.

With the convening of the course, students are given presentations by the partners and begin the initial stages of problem definition, which will soon include a more precise interviewing of the partner for technical information and the formal preparation of a problem statement and a work plan. At this time the student teams launch themselves, define roles, and set a schedule for themselves and the partner. They will also establish their preliminary metrics for success, based on technical and EHS performance standards and the needs of the client.

The teams develop preliminary strategy options based on an investigation of possible alternatives inspired by instruction in and examples from systems, design, and bio-inspired thinking. In this stage they are encouraged to be divergent in their approach, keeping function as the key touchstone in a wide-ranging search for solutions. Moreover, they are encouraged to "stay in the problem space" and not rush to any investment in one solution. Templates are provided to prompt students to look at as many facets as possible in solving the problem. Can we achieve the same results through manipulation of material, space, or time? for instance. Do we need more information, more energy, or a different type of organization, in order to solve our challenge? The emphasis is on intervening in a system, rather than searching for a replacement item, and eliminating the inherent hazard of chemical use, rather than mitigating its risk or effect.

Next, teams will assess the prospects for their different options, and switch to a convergent approach. They will typically take one such strategy and "road test" it through the entire course process, from setting performance criteria to performing a chemical hazard analysis to analyzing, and presenting results. This gives teams a hands-on understanding of challenges and expectations associated with the final report, the Opportunity Map. This is also a crucial stage where teams discover for themselves the opportunities and constraints associated with the challenge and with any options that they have chosen to pursue. The review and critique of their "road test" results are one of the more important milestones of the course and the entire class, instructors, and often guest professionals, participate in this exercise.

The comparison of technical and EHS performance standards, first in the "road test" and then in the final Opportunity Map is really the heart of the course, although a great deal of work is done to prepare teams for this stage. This is, of course, an alternatives assessment in its basic sense, but customized to the academic goals of the course. For the chemical hazard assessment students develop their search in phases, first going to curated authoritative lists, then individual topic literature searches, and then predictive tools as they strive to fill the numerous data gaps that exist. They start their search at the Chemical Hazard Data Commons of the Healthy Building Network and work their way deeper into the leads that they find there. They are guided by the criteria outlined in the Green Screen for Safer Chemicals developed by Clean Production Action but do not perform a complete version of this assessment.

Finally, the student teams apply their methodology to their most promising options, comparing them within the Green Screen framework, technical criteria agreed upon with the partner, and any other rules that had become necessary during their search. They rank them according to metrics that they had devised, and synthesize their results, including sustainability context, relative technical feasibility, and relative confidence level in the data results and sources. They are encouraged to present actionable recommendations to the partner and to characterize all their recommendations on a scale of relative feasibility, specifically tailored to the partner as well as the industry sector.

With a first draft of the Opportunity Map complete, the teams hone their communication skills and then present their findings to the partner and public. After gaining feedback from this, they complete their final drafts and submit them. Often, individual students continue the work started in the course through support from BCGC: presenting at conferences, working as interns and testing concepts in the laboratory, and being hired by partners or other stakeholders.

Foundational content

This course is necessarily interdisciplinary and is taught by a diverse team of instructors who also act as coaches for the teams. The following are the subjects that are instructed formally as support to and preparation for the team project effort. Three instructors in Chemistry, Public Health, and Innovation coach teams throughout the project, while guest instructors contribute the other content as needed.

Chemistry: This graduate course benefits from the resources of one of the premier colleges of chemistry in the world and draws typically from a pool of highly competent doctoral chemistry students. Project teams are assured of deep chemistry expertise within their ranks, but the experience of the "resident" chemist(s) may not be specifically aligned with the current challenge and students may not be aware of some of the basic tenets of green chemistry and its relationship to toxicology and public health. Current chemistry education for most students has not included much of this content. Students, therefore, receive early lectures on the principles of green chemistry, and the specific area of inquiry, for example, organic chemistry as it may relate to polymers. They also receive later lectures on the chemistry of some of the biological processes introduced in the bio-inspired design sessions. Throughout the course, they have access to a chemistry instructor.

Public Health: This course was cofounded by a physician who continues to teach the course, and is an elective in the Industrial Hygiene master's degree program. Human and environmental health, therefore, are of paramount concern in the educational and outreach objectives of the course. Students receive initial lectures in the state of public health related to chemical production, use, and disposal, as well as training in research and performance of a comparative chemical hazard assessment.

Toxicology: Many of the public health students in the course may have had some toxicological training and education, but this material is new to almost all chemistry and engineering students. They, therefore, receive an introductory lecture on toxicological pathways and toxicology practice and terms, and then later are instructed in the use of a computational toxicology tool when data gaps in team findings have been identified.

Law and Policy: This course was designed as a comprehensive problem-solving and innovation course, and students are given as wide background in a chemical challenge as possible within the 16 weeks of a semester. This includes a lecture on the state of play in pertinent chemical regulations and policy, typically delivered by a guest instructor from the School of Law, UC Berkeley.

Business: While cost-benefit analysis, marketing, and business plan making are not part of this course, the concerns and drivers of behavior within the industry, government, or nonprofit sector are important contextual issues that student teams need to be aware of in order to provide actionable recommendations. The challenge partners typically impart this intelligence in their initial introductory lectures, and student teams have a subsequent opportunity to query the partners in a formal interview and review their problem statement (or design brief). In some cases, a guest instructor has been invited to lecture on the business perspective related to the challenge.

Innovation: Systems, Design, and Bio-inspired Design Thinking: The design and creative thinking part of the course is introduced at the start of the course, as it is essential to the framing of the problem and the early generation of ideas. Students receive lectures on systems thinking and design thinking and are shown examples of principles of biologically inspired design, and case studies of this translation process. They then practice the methodology in a sample exercise aligned with their final challenge statement.

Professional Practice: skills, as well as content, are transferred during this course. Chief among them are research methods for performing a comparative chemical hazard assessment; project and team management; client relations, particularly preparing a problem statement, interviewing a client, and conducting professional work sessions; and written, oral and graphic communication.

Key sources

One of the main research activities of the Greener Solutions teams is the conducting of the comparative chemical hazard assessment and the information needed is gleaned from several sources, typically in a step-wise fashion. Team members typically start with a review of curated lists for toxicological information about candidates, move on to a more in-depth investigation into select peer-reviewed journals for both technical and EHS information, and then finally to the use of surrogate indicators, molecular structure comparison, and computational toxicology to make judgments where data gaps continue to exist.

Students are encouraged to design their own ranking logic based on the insights collected from the challenge partner and the available literature, but all of the rankings have been generally based on the Globally Harmonized System (GHS) of toxicological endpoints, and assessments are similar to the approach taken in GreenScreen. An important initial information source is the Chemical Hazard Data Commons of the Healthy Building Network.

Outcomes

Desired outcomes from the course can be divided into education, professional development, and engagement with the wider community. Educational outcomes can be further divided into content and skills development. Students should complete the course with an understanding of the principles of green chemistry, the relationship of material design to public health, and the ability to conduct a comparative chemical hazard assessment. In addition, they should have a greater familiarity with systems and design thinking, interdisciplinary, team-based problem-solving, and project management. The course has also been an important bridge between the university and the wider community of organizations and individual professionals. This has led to several collaborations that have spanned many years. Some of these collaborations are explained further in the case studies to follow.

The professional development of the next generation of public health, chemistry, and engineering professionals is perhaps harder to measure than the adoption of the resultant technical innovations but is equally important as fulfilling the academic and public health outreach outcomes of the course. The program has had an important and sometimes pivotal impact on the careers of many participating individuals. Affiliated students have gone on to academic positions, government agencies, the nonprofit world, and private industry and carried their Greener Solutions experience with them.

Further reading

ACS. https://www.acs.org/content/acs/en/chemical-safety/basics/ghs.html.

ACS. https://www.acs.org/content/acs/en/greenchemistry/about.html.

ACS. https://www.acs.org/content/acs/en/meetings/greenchemistryconferences.html.

Berkeley Center for Green Chemistry. http://live-bcgc.pantheon.berkeley.edu/greenchemconf.

Berkeley Center for Green Chemistry. https://bcgc.berkeley.edu.

Berkeley Science Review. https://berkeleysciencereview.com/2011/04/a-fresh-look-at-green-chemistry/.

Chemistry. https://chemistry.berkeley.edu/home.

Department of Toxic Substances Control. https://dtsc.ca.gov/grsp/.

Department of Toxic Substances Control. https://dtsc.ca.gov.

Ellen MacArthur Foundation. https://www.ellenmacarthurfoundation.org/.

Environmental Health Hazard Assessment (OEHHA). https://oehha.ca.gov.

Environmental Health Hazard Assessment (OEHHA). https://oehha.ca.gov/public-information/events/public-forum-identifying-and-evaluating-alternatives-case-bpa-free-can.

EPA. https://www.epa.gov/p2.

FDA. https://www.fda.gov/.

Green Chemistry & Commerce Council. https://greenchemistryandcommerce.org/.

IGERT. http://www.igert.org/public/about.html.

Mulvihill, M. J., Beach, E. S., Zimmerman, J. B., & Anastas, P. T. (2011). Green chemistry and green engineering: A framework for sustainable technology development. *Annual Review of Environment and Resources, 36,* 271−293.

NIEHS. https://www.niehs.nih.gov/research/atniehs/dntp/index.cfm.

Northwest Green Chemistry. https://www.northwestgreenchemistry.org/.

Pharos Project. https://pharosproject.net/.

Philomathia. https://www.philomathia.org/green-chemistry-collaborative-approaches-new-solutions-march-24-2011/.

Safer Alternatives. https://www.saferalternatives.org/about.

TU Delft. https://www.tudelft.nl/en/.

Turi. https://www.turi.org/Our_Work/Alternatives_Assessment/Alternatives_Assessment/Tools_and_Methods.

Schwarzman, M. R., & Buckley, H. L. (2019). Not just an academic exercise: Systems thinking applied to designing safer alternatives. *Journal of Chemical Education, 96*(12), 2984−2992.

UNECE. https://unece.org/DAM/trans/danger/publi/ghs/ghs_rev04/English/ST-SG-AC10-30-Rev4e.pdf.

Vegosen, L., & Martin, T. M. (2020). An automated framework for compiling and integrating chemical hazard data. *Clean Technologies and Environmental Policy, 22*(2), 441−458.

York. https://www.york.ac.uk/chemistry/research/green/.

Chapter 3

Green Chemistry case study on preservatives: replacing Phenoxyethanol and Isothiazolinones in personal care and household products

What was the challenge

The overall challenge was to develop cheap, safe, and effective antimicrobial preservatives for use in cleaning and personal care products that would replace currently used chemicals of concern like Phenoxyethanol and Isothiazolinones.

The USDA Agricultural Research Service (ARS), Western Regional Research Center in Albany, California, the personal care product manufacturer Method, and the Berkeley Center for Green Chemistry at the University of California, Berkeley conducted collaborative antimicrobial research and development from 2014 through 2019. The work has gone through four phases leading to the current solutions described below and is ongoing at this writing.

Phase 1 comprised a semester-long investigation of personal care product preservative alternatives by the graduate public health course, Greener Solutions, in the fall, of 2014. This was in partnership with the companies Method, Seventh Generation, and Beautycounter.

The Greener Solutions team took a novel and nontraditional approach to the challenge, starting with functional performance and then looking for mechanisms of microbial control found in the biological world. They then classified four groups of functional compounds found in nature based on a review of the scientific literature and chose one group with the most theoretical promise for meeting the physiochemical requirements of the commercial sector.

During Phase 2 of the project Dr. William Hart–Cooper coordinated with Dr. Kaj Johnson, a product developer at Method, with support from researchers at the USDA-ARS and the Berkeley Center for Green Chemistry. Hart–Cooper and Johnson established experimental protocols for screening and

Green Chemistry in Practice. https://doi.org/10.1016/B978-0-12-819674-8.00002-3

51

testing target antimicrobials. Over 200 candidate preservative compounds were tested and a range of hazard assessments and computational toxicology analysis of candidates was completed.

Early research in this second phase indicated that hydroxybenzoate compounds were more promising candidates than terpenes, and further hazard profiling yielded octyl gallate as the leading compound. This research was published in 2017, in the ACS Journal Sustainable Chemistry and Engineering under the title "Design and Testing of Safer, More Effective Preservatives for Consumer Products."

Phase 3 of this project comprised research and documentation for submittal to a national industry competition for an alternative preservatives design for the household and personal care sectors. As part of a cooperative research and development program, USDA and Method developed a new class of two-subunit compounds that decomposed rapidly when their function was no longer necessary. These "reversible" preservatives could produce their desired preservative function when bound together as they would be in a consumer product but would degrade to harmless subunits once they entered the environment. The Berkeley Center for Green Chemistry supported a hazard analysis of these substances.

Phase 4 entailed extensive benchtop testing of formulations per commercial standards by staff at USDA-ARS, and the development of a potentially disruptive paradigm for antimicrobials employing "reversible bonding" in a safer, broad-spectrum preservative. This approach and documentation secured a first prize for the research team in the Green Chemistry and Commerce Council (GC3) Preservative Challenge and formed the foundation for a patent application. The ARS team has subsequently applied for a US patent.

Why this project was important

Antimicrobials are vital components of many consumer products, especially personal care products, where they are used to prevent microbial contamination that would reduce shelf life, product appeal, and utility. In some cases, however, these compounds could present a danger to consumers. Since antimicrobials are bioactive by design, the chemistry that makes them useful can also make them dangerous to both human and environmental health. Most commonly, they cause skin, eye, and respiratory irritation in humans, but they can also have chronic systemic effects such as cancer and reproductive toxicity.

A 1980 report by the US EPA, Sources of Toxic Compounds in Household Wastewater, found that consumer cleaning and personal care products were responsible for the bulk of preservative chemicals found in residential wastewater. Widespread dissemination of antimicrobial chemicals can lead to antimicrobial resistance and aquatic toxicity

Antimicrobials can be described as substances that prevent, kill, or reduce the growth of microbes such as fungi and molds, bacteria, viruses, and protozoa (generally termed "germs"). The U.S. EPA regulates disinfectants that are used on environmental surfaces (housekeeping and clinical contact surfaces) by the Antimicrobials Division, Office of Pesticide Programs, EPA, under the authority of the Federal Insecticide, Fungicide, and Rodenticide Act (FIFRA) of 1947, as amended in 1996. Under FIFRA, any substance or mixture of substances intended to prevent, destroy, repel, or mitigate any pest, including microorganisms but excluding those in or on living man or animals, must be registered before sale or distribution. To obtain a registration, a manufacturer must submit specific data regarding the safety and effectiveness of each product. More than 4,000 antimicrobial pesticide products are currently registered with EPA and sold in the marketplace. These products are found in the following types of products: personal care, soaps, paints, exercise mats, apparel, food storage containers, textiles, keypads, kitchenware, school supplies, and countertops.

When used in personal care products like shampoos and moisturizing creams, antimicrobials are regulated by the Food and Drug Administration (FDA) and the Federal Food, Drug, and Cosmetics Act (FDCA).The FDA also regulates cosmetics under the authority of the Fair Packaging and Labeling Act (FPLA). The FDCA regulations are germane when a product is adulterated or it is misbranded. "Adulteration" refers to violations involving product composition–whether they result from ingredients, contaminants, processing, packaging, or shipping and handling. "Misbranded" refers to deceptively packaged or mislabeled products., including the lack of directions for safe use or warnings.

The FDA does not test or approve products before market entry and re-quires the registration of cosmetic companies or their formulas. Testing and improvement of formulations for safety and health are, therefore, largely the responsibility of the private manufacturing sector. The FDA does provide guidance documents and monitors recall of products.

Viewed from the standpoint of risk, cosmetics, and household products can be especially impactful and of concern when containing harmful ingredients. The formula for risk is typically a combination of inherent chemical hazard, exposure to the substance, and the relative vulnerability of the population being exposed. Rates of exposure can be high for some products like skin cream, as they are applied directly to the skin every day. Other home or personal products might also particularly affect vulnerable populations, like infants or young children.

Because of the oftentimes direct and intimate contact of consumers with these products, they elicit more regular concern and direct engagement with the companies providing them. Personal health concerns consistently rank high in priority for consumers when asked about chemical safety. Many companies have responded to these concerns and indeed some have estab-lished their brand distinction based on safety and health. Seventh Generation, Method, and BeautyCounter, partners in the initial study described here, have led their industry sectors in the pursuit of safer products. Below are two examples of the type of antimicrobial that these companies were hoping to replace.

Phenoxyethanol is a broad-spectrum preservative with effectiveness against a wide range of Gram-negative and Gram-positive bacteria, yeast, and mold. It is also used as a solvent and, because of its properties as a solvent, it is used in many blends and mixtures with other preservatives. It is used in cosmetics and also as a stabilizer in soaps and perfumes. It is found in a wide variety of personal care products: makeup, shampoo, sunscreen, body wash, hair products, shaving cream, baby wipes, soap, deodorant, and toothpaste, among others. It has been linked to eczema and allergic reactions. Infant oral exposure to phenoxyethanol can acutely affect the nervous system function. Phenoxyethanol is also of concern from an environmental perspective and has been found in Japanese rivers in concentrations normally associated with consumer product formulations, due to high levels of cosmetics and household detergents in wastewater discharge.

The FDA had found phenoxyethanol to be deemed safe when used topically in concentrations less than 1% in a 1990, Cosmetic Ingredient Review (CIR) and in a subsequent 2007 update. This chemical is restricted in use in Japan, and in the European Union. The European Economic Community (EEC) Cosmetics Derivative and the Cosmetics Regulation of the European Union has approved phenoxyethanol in concentrations up to 1%.

Isothiazolinones: methylisothiazolinone (MIT) and benzisothiazolinone (BIT) are biocidal and used as preservatives in a range of household products, like laundry and dish detergent. These molecules are skin sensitizers, skin and eye irritants, and exhibit acute aquatic toxicity, so Seventh Generation had made the elimination of MIT and BIT from their products a priority. These products included dish liquid, dish gel, laundry detergent, fabric stain remover, and fabric softener.

Methylisothiazolinone was banned in body lotions, deodorants, and other leave-on cosmetics sold in the European Union in 2016, and the European Commission's Standing Committee on Cosmetic Products limited the amount of methylisothiazolinone in rinse-off personal care products to 15 ppm in 2017.

Project phases and solutions

An interdisciplinary approach was necessary to achieve a balance of technical and EHS performance as demanded by the commercial market. The new preservatives needed to be effective and to have a favorable life cycle, and human health, and EHS profiles. Success required expertise in a variety of research fields that would each play a role over the several phases of this project: synthetic and environmental chemistry, bio-inspired design, microbiology, human and environmental toxicology, product formulation, and consumer regulatory standards. Moreover, progress toward a solution evolved in phases with different participants and emphasis, depending on the discoveries made previously.

Phase 1: The Greener Solutions course: finding four classes of antimicrobials from nature, 2014

The stated challenge was to provide recommendations for safer antimicrobial preservatives in the home and personal care products market. It required learning about the industry sector, the current chemicals of concern used in formulas (including paraben esters, formaldehyde (or formaldehyde releasers), isothiazolinones, dichlorophene, and iodopropynyl butylcarbamates, 1–3) and searching for more benign functional substitutes.

To meet these needs during the first phase of the project, students and instructors in the Greener Solutions course consulted with partners at Beautycounter, the Method Corporation, Seventh Generation, the USDA Agricultural Research Service in Albany California, the Department of Civil and Environmental Engineering at the University of California Berkeley, and the Center for Human and Environmental Toxicology at the University of Florida. The Greener Solutions student team was composed of Billy Hart–Cooper, Heather Buckley, Adam Byrne, and Jiawen Liao, and they worked with Mia Davis from Beautycounter; Clement Choy, Chantal Bergeron, and Martin Wolf at Seventh Generation, Mark Dorfmann at Biomimicry 3.8, and Kaj Johnson and Ryan Williams at Method. Additional research support was provided by Larry Weiss of AoBiome.

The Greener Solutions group used bio-inspired design thinking to create a list of chemicals that commonly performed preservative functions in nature and also met the safety and performance standards required by the client groups. Then the team proposed additional compounds based on their initial findings and determined that the compounds on their lists could be organized into four classes of compounds: terpenes, peptides, flavonoids, and lipids. These types of chemicals occur naturally in pine trees, milk, tea leaves, and fish oil respectively.

The team assessed individual chemicals for antimicrobial efficacy by reporting minimum inhibitory concentration (MIC). They identified antimicrobial candidate chemicals exhibiting MICs that were comparable or superior to current industry standards. These compounds were found to be effective against a broad spectrum of microbes, including Gram-positive and Gram-negative bacteria, yeast, and mold. The team then completed hazard assessments that categorized these chemical classes by 18 acute and chronic health endpoints, noting data gaps (Please see Table 3.1). This chemical hazard assessment was modeled directly from Clean Production Action's GreenScreen for Safer Chemicals and compared the four classes of natural alternatives to two commonly used compounds, phenoxyethanol, and Methyliso-thiazolinone.

TABLE 3.1 Summary of hazard assessment of proposed green preservatives and industry standards.

| | Human Health Group I | | | | | Human Health Group II | | | | | | | Environmental Health | | Environmental Fate | | Physical Hazards | |
| | Carcinogenicity | Mutagenicity & Genotoxicity | Reproductive Toxicity | Developmental Toxicity | Endocrine Activity | Acute Mammalian Toxicity | Systemic Toxicity & Organ Effects | Neurotoxicity | Skin Sensitization | Respiratory Sensitization | Skin Irritation | Eye Irritation | Acute Aquatic Toxicity | Chronic Aquatic Toxicity | Persistence | Bioaccumulation | Reactivity | Flammability |
	C	M	R	D	E	AT	ST	N	SnS	SnR	IrS	IrE	AA	CA	P	B	Rx	F
Terpenes	DG	M*	DG	DG	DG	M	DG	DG	M	M	H	H	H	H	M	L	L	M
Peptides	DG	DG	DG	DG	DG	L*	M*	DG	DG	L*	DG	DG	L*	L*	L*	L	L	L
Flavonoids	DG	M	DG	M	H	M*	DG	DG	M*	L	M*	M*	DG	DG	M	L	L	L
Lipids	M*	DG	DG	DG	DG	L*	DG	DG	M*/L*	L*	H	H/L*	H	M	L	L	L	L
Phenoxyethanol	DG	L	M	M	DG	M	DG	DG	M*	DG	M*	H	L	M	L	L	L	M
Methylisothiazolinone	M*	L	M*	M*	DG	M	M	DG	H	H	H	H	H	H	M	L	L	L

From Buckley, H., Byrne, A., Hart-Cooper, W., & Liao, J. (2014). *Next generation chemical preservatives: Protecting People, products, and our Planet. Unpublished report* available at: https://bcgc.berkeley.edu/greener-solutions-2014/.

These natural products and their derivatives represented drop-in replacements for existing compounds conforming to current health and efficacy standards and had desirable physicochemical properties such as hydrophobicity and (in most cases) low volatility, which would make them appealing to formulators.

Each of the different chemical classes demonstrated strengths and weaknesses. Among the terpenes, those with polar groups were generally most effective, especially when the structure included alcohols and phenols. Terpene compounds were highly effective at low concentrations but did present the highest potential for skin irritation and sensitization at high concentrations.

Peptide compounds, in contrast, had very favorable hazard profiles, but less favorable antimicrobial efficacy. Phosvitin, a component of egg yolk that has been studied as an antimicrobial and antioxidant, demonstrated some potential as a commercial antimicrobial, as did polylysine peptides. Antimicrobial peptides were noted to be common in nature, particularly antibacterial peptides, but the team found that these peptide-based toxins were generally not effective against fungi—which were often the organisms producing them. The biggest obstacle to the adoption of antimicrobial peptides, however, was the industrial and logistical challenges: producing antimicrobial peptides in bulk would be costly, and the residues would be very sensitive to pH and peptidase compounds, which are often contained in product formulations.

Flavonoids presented the third class of potential antimicrobials, given their noted efficacy in plants and frequent appearance in teas and spices. The students found that the mechanisms of action for flavonoids were incompletely understood, and that animal studies reported that high oral dosing led to reproductive toxicity and endocrine disruption, despite the apparent safety of these compounds in the context of the foods. Capric, lauric, myristoleic, and palmitoleic acids were identified as effective antimicrobial agents, but not against gram-negative bacteria. Also, there was significant concern about aquatic toxicity and possibly chronic exposure toxicity in humans, since these compounds were known to enhance the uptake of other substances through the dermal exposure route.

The proposed substitutes compared favorably with typical industry antimicrobial compounds. All of these solutions had more favorable hazard profiles than methylisothiazolinone, and all except for terpenes had better hazard profiles than phenoxyethanol, but the group reasoned that terpenes could be used at much lower concentrations than phenoxyethanol, reducing the hazard. Therefore, of these classes, the team determined that terpenes balanced production feasibility, safety, and efficacy most effectively, and so they were prioritized for additional investigation during the second phase of the project: the BCGC-USDA-Method partnership.

Phase 2: Continued research and a publication, 2015−16

During the second phase of the project Dr. Billy Hart−Cooper, supported by a grant from BCGC, and Dr. Heather Buckley coordinated with product developer Johnson at Method and researchers at the USDA-ARS that included Jong H. Kim, Luisa W. Cheng, Kathleen L. Chan, and William J. Orts, to conduct a systematic analysis of candidate preservative compounds. Once the team identified and synthesized the compounds of interest, they conducted antimicrobial efficacy studies and a range of hazard assessments. BCGC's SAGE program fellow David Faulkner assisted with the human health hazard analysis of the compounds and provided computational toxicology analysis to make toxicity predictions when data gaps arose. Drs. Susan Amrose, at the UC Berkeley Civil and Environmental Engineering Department, Christopher Vulpe at the University of Florida Center for Human and Environmental Toxicology, and Martin Mulvihill, at Safer Made provided additional research support (Fig. 3.1).

Early research in the second phase of the project indicated that hydroxybenzoate compounds were more promising candidates than terpenes: hydroxybenzoates were more readily synthesized (or commercially available), more likely to be product compatible due to their amphiphilicity, and could be "tuned" for solubility by adjusting the length of their alkyl chain. Jong Kim at the USDA was particularly important to these early decisions. After synthesizing and testing a series of gallate compounds, the team observed a trend in the structure-activity relationships between changes in polarity (as determined by the length of the hydrocarbon chain on the hydroxybenzoate) and antimicrobial activity. They found the optimal carbon chain length to be around seven or eight, and so octyl gallate was the subject of additional hazard analysis in addition to further efficacy testing. As expected, octyl gallate was the most effective of the test compounds and performed better in the fungal and bacterial growth tests than several preservatives in current commercial use, including methylparaben, phenoxyethanol, sorbic acid, and benzoic acid.

A computational "test battery" was devised to make use of several software tools (to account for biases in the datasets used to develop the predictive tools),

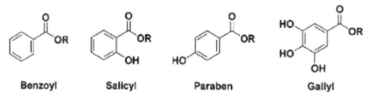

FIGURE 3.1 General structures of benzoates, salicylates (2-hydrox- ybenzoates), parabens (4-hydroxybenzoates), and gallates (3,4,5- trihydroxybenzoates). *From Buckley, H. L., et al. (2017). Design and testing of safer, more effective preservatives for consumer products. ACS Sustainable Chemistry and Engineering, 5(5), 4320−4331.*

and as a result, greater confidence could be applied to the overall prediction. The testing schema compared the novel gallate derivatives to contemporary preservative compounds commonly used in consumer products, which provided benchmark levels for efficacy and hazard endpoints. Compared to the other test compounds, octyl gallate had a more favorable hazard profile and, compared to contemporary commercial preservatives, it generally had a comparable hazard profile.

In Kim (2016), Microbiology Discovery published Kim et al., "Antifungal efficacy of octylgallate and 4-isopropyl-3-methylphenol for control of *Aspergillus*". This work compared 21 generally regarded as safe (GRAS) compounds with six conventional preservatives in the control of the fungus *Aspergillus*. Octylgallate (OG) showed the highest antifungal activity. The group also investigated synergism between OG and conventional preservatives and found that the efficacy of 4-isopropyl-3-methylphenol (4I3M) was augmented four-fold.

In 2017, the American Chemical Society published Buckley et al., "Design and testing of safer, more effective preservatives for consumer products" in Sustainable Chemistry and Engineering. This article compared several phenolic preservative alternatives to the common preservatives found in personal care products and building materials. The team quantified antimicrobial activity against *Aspergillus brasiliensis* (mold) and *Pseudomonas aeruginosa* (Gram-negative bacteria), and conducted a hazard assessment, complemented by computational modeling, to evaluate the human and environmental health impacts of these chemicals. Their written findings showed that octyl gallate demonstrated better antimicrobial activity and comparable or lower hazards, compared to those preservatives currently in use.

Phase 3: A change in approach for a national competition: investigating reversible-bond heterodimers, 2017

In April 2017, the Green Chemistry and Commerce Council (GC3) announced a new Challenge, "Developing New Preservatives for Personal Care and Household Products:" a competition to design an environmentally-friendly chemical agent with broad spectrum efficacy against bacteria, mold, and yeasts in personal care products. USDA-ARS, supported by BCGC, worked to build on the previous preservatives work and submitted an entry to the competition in late August 2017. The eventual development approach, however, would be in a new direction.

The USDA-ARS team comprised Jong H. Kim, William M. Hart—Cooper, Luisa W. Cheng, Kathleen L. Chan, Diana Franqui, Lauren Lynn, and William J. Orts. Kaj Johnson from Method continued to provide his insights on formulation and industry methods and materials. The BCGC support group comprised former executive director Martin Mulvihill, postdoctoral student Heather Buckley, and doctoral candidate David Faulkner.

Although octyl gallate had been demonstrated to be effective, its environmental health profile and formula compatibility were not satisfactory. Dr. Hart—Cooper, now working full-time at ARS, met with Kaj Johnson of Method and his supervisor at ARS, William Orts, and proposed a new approach to the challenge: a new class of two-subunit compounds that decomposed rapidly when their function was no longer necessary.

The heterodimer consisted of an aldehyde subunit and an aminoguanidine subunit, which would remain in an active complex when present in high enough concentrations, but which would dissociate upon dilution. These "reversible" preservatives could produce their desired preservative function when bound together as they would be in a consumer product, but would degrade to harmless subunits once they entered the environment.

A series of compounds based on the phenolic aldehydes and cinnamaldehyde were also synthesized and run through the same battery of tests for bacterial and fungal contamination, environmental fate, and human health endpoints as the gallate-based preservatives. These compounds were chosen based on their commercial availability, established human health profiles, and physicochemical properties. *trans*-cinnamaldehyde and 2-hydroxy-5-methoxybenzaldehyde (commonly known as vanillin) emerged as the most efficacious compounds. These compounds had favorable human and EHS evaluations, and where they demonstrated some concern, they were found to be less hazardous than existing preservatives. Dr. David Faulkner again provided insight into the toxicological hazard and computational testing as needed to fill data gaps using the same computational test battery as before.

Phase 4: Performance testing of a novel solution, winning of a national prize and patent application, 2017, 2018

The ARS/Method team prepared a class of broad-spectrum, reversible guanylhydrazone antimicrobials derived from aminoguanidine and aldehydes sourced from food-grade cumin seed oil. Guanylhydrazones had an extensive record in biomedical applications and their hydrazone bond reversibility was well documented. Guanylhydrazones are thermodynamically stable when in use but can break down quickly when subjected to external stimuli like dilution and/or a pH change. Thus the team had an approach of a kinetically dynamic compound comprised of relatively safe components that could break down quickly when disposed of.

Previous studies had indicated that guanylhydrazones were inhibitory to classes of bacteria like *Pseudomonas* and *Staphylococcus*, so the team was encouraged to begin testing their candidate formulas as preservatives and disinfectants. They prepared 17 variations of guanylhydrazones and tested them against conventional antibiotics like isothialzolinones for efficacy against mold, bacteria, and yeast, and found them to be comparable, with one formulation being one to two orders of magnitude more potent.

They also tested these formulas against the conventional disinfectants, bleach, and benzalkonium chloride and found them to be comparable.

As part of the GC3 competition, the team tested the durability of two of these antimicrobial formulas under the stress of environmental conditions such as anionic surfactants and included their preservative within commercially formulated spray cleaners containing them. All detectable yeast and bacteria and nearly all the mold were eliminated within the standards of the USP Chapter 51 Preservative Challenge Test. Moreover, the team conducted tests for biodegradation of aminoguanidine HCl after the end of use by mixing the formulas with wastewater and measured a half-life of 32 days. In contrast to conventional antimicrobials, the constituents of one of the formulas have been shown to biodegrade readily and are up to 2600 times less acutely hazardous to aquatic life than conventional preservatives and disinfectants.

Several commercially-viable self-assembled compounds passed testing criteria for commercial applications at concentrations that were a fraction of the levels typically required for green, "safe" antimicrobials (0.2% vs. 1% −2%). The active levels and performance were comparable to industry standard formulations of isothiazolinones, formaldehyde releasers, and quaternary ammonium compounds, all of which exhibited severe ecotoxicity and antimicrobial resistance concerns.

This strategy was novel for antimicrobial agents and represented a paradigm shift in the development of biocides and safer chemicals in general. This invention is the basis for developing reversible actives in other chemical classes of concern. The team's presentation on the technology was awarded first place in the Green Chemistry and Commerce (GC3) Preservative Competition and was subsequently highlighted in Chemical and Engineering News ("The search is on for new cosmetic preservatives", 2018) (Fig. 3.2).

The reversible antimicrobials produced were coinvented by scientists at Method and USDA as part of a cooperative research and development agreement funded by both parties. The Berkeley Center for Green Chemistry provided funding, chemistry research, and analysis toward safer antimicrobials and toxicological analysis of the reversible antimicrobials.

An international patent, "Self-assembled Active Agents", was applied for on March 5, 2018 and published on September 13, 2018. It is viewable on https://patentscope.wipo.int/search/en/detail.jsf?docId=WO2018164994

The following is its abstract:

A self-assembled active agent may be formed by a process including covalently bonding at least a first component molecule and a second component molecule, the two-component molecules displaying synergy such that the effective amount of the self-assembled active agent is lower than the sum of the effective amounts of the first component molecule and the second component molecule. The component molecules may be chosen such that the covalent bonding is reversible, for example through a hydrazone bond between an amine and an aldehyde.

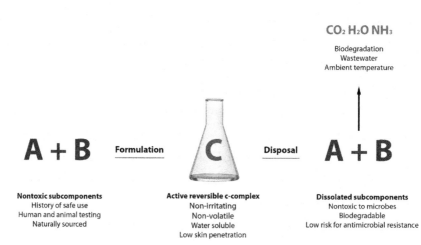

FIGURE 3.2   Schematic of a reversible antimicrobial agent's lifespan. *Adapted from Hart-cooper, W. M., Orts, W. J., Johnson, K., Lynn, L. E., Franqui-villanueva D. M. (2018). Self-assembled active agents. Available at: http://www.freepatentsonline.com/y2018/0303100.html.*

The active agent may thus have a controllable activity such as an antimicrobial agent, a biocide, an antiviral agent, a preservative, an antifouling agent, a disinfectant, or a sensor agent, such as for a particular molecule or for pH.

What was innovative about the solution

This research work has produced a nonirritating, low eco-toxic, broad-spectrum preservative booster made from two nontoxic components. When diluted in water treatment plants, the preservative molecule dissociates into benign biodegradable components, This is a new paradigm in the creation of biocides and safer chemicals in consumer products. The design of reversible antimicrobials constructed from biodegradable and renewable subcomponents has never been used previously in this industrial sector and makes two contributions to the movement toward safer chemicals in products. It provides both a new substitute ingredient for a range of household and personal care products that is orders of magnitude less toxic than current chemistries and a conceptual model for similar investigations for other classes of chemicals.

Reversible antimicrobials are effective biocides in concentrations applicable to their functional performance while in personal care and household products but break down into benign subcomponents when diluted with water. This ability has the potential to reduce or eliminate a significant portion of the volume of antimicrobials currently found in the nation's waterways.

What was the impact

Reversible antimicrobials as described in this chapter offer an entirely novel class of compounds available for both disinfectant and preservative functions to a range of industrial markets typically estimated in the billions of dollars annually. These markets include but are not limited to personal care products, household products, and coatings. The use of this solution type could potentially address the following serious issues: skin sensitivity outbreaks, and growing microbe resistance (so-called "superbugs"). Moreover, the chemical design concept of reversible bonding offers the prospect of a wider choice and greater efficacy of products, and the ability to tune the toxicity of products per the environment in order to reduce risk.

Beyond the human health benefits, widespread adoption of these safer, reversible preservatives could replace aquatically toxic chemicals and substantively reduce the environmental hazard of household, health, and beauty products by effectively deactivating them before residential wastewater discharge. Their intentional design for dissociation and biodegradation should reduce collateral ecotoxicity and the development of antibiotic resistance. Early reformulation of some of Method's products, for example, resulted in the avoidance of an estimated 40,000 lbs. of hazardous preservatives and one million lbs. of hazardous surfactants.

There are myriad potential applications for this new class of reversible preservatives in the Health and Beauty and Household Cleaners market sectors, and likely unforeseen uses in other sectors as well. The technology that allows the preservative to function while contained within the product formulation—but not outside of it—represents a paradigm shift in the way that formulators think about nonactive ingredients in consumer products. Beyond preservatives, other common additives could be engineered to be "reversible," exuding their effects while in the product formulation, and degrading to harmless biomolecules once diluted in the environment.

The Health and Beauty sector alone is a large and expanding market. From 2013 to 2017, the US market for Health and Beauty personal care products grew from $75.78 billion USD to $86.134 billion USD, and the global cosmetics market alone is anticipated to be worth $805 billion USD by 2023. A mere five corporations account for over 90% of the US market for beauty products, so if any one of them were to adopt reversible preservatives, the reduction in human and environmental hazards could be extensive. This is an "upstream" value chain intervention that occurs at the formulation stage and thus obviates the need for protection and management costs associated with risk and remediation and environmental cleanup.

This technology adds a temporal aspect to the concept of "tunable" molecules, providing chemists with another tool to design compounds that can be programmed to automatically degrade across different timescales without the aid of enzymes or other reactions—the degradation is a structural feature of these compounds that activates according to the laws of thermodynamics. Already, the concept of thermodynamically-driven auto-degradation has entered the fields of molecular biology and drug development, where it has potential drug-delivery applications. Widening this strategy to consumer products would do much to improve the health and safety of citizens and the environment.

As with many of the projects initiated at the Berkeley Center for Green Chemistry, the development of reversible preservatives has spanned several years and involved changing research approaches, collaborative mechanisms, and the growth of new leaders. This project started as a literature search and analysis in a multi-disciplinary graduate course with industry partners and segued into a funded internship for laboratory testing of the ideas generated in the course. This was enabled by a cooperative research and development agreement between the company and government partners. Further research and innovation were sparked by a national competition hosted by a green chemistry trade association. Most recently it is being pursued as part of the missions of a government agency and a commercial company with patents and commercial development on the horizon. Each phase, despite changing participants, leaders, and working arrangements, has contributed significantly to the progress outlined here.

References

Buckley, H., Byrne, A., Hart-Cooper, W., & Liao, J. (2014). *Next generation chemical preservatives: Protecting People, products, and our Planet.* Unpublished report available at: https://bcgc.berkeley.edu/greener-solutions-2014/.

Buckley, H. L., Hart-Cooper, W. M., Kim, J. H., Faulkner, D. M., Cheng, L. W., Chan, K. L., & Mulvihill, M. J. (2017). Design and testing of safer, more effective preservatives for consumer products. *Acs Sustainable Chemistry & Engineering, 5*(5), 4320−4331.

Hart-cooper, W. M., Orts, W. J., Johnson, K., Lynn, L. E., & Franqui-villanueva, D. M. (2018). *Self-assembled active agents.* Available at: http://www.freepatentsonline.com/y2018/0303100.html.

Kim, Jong H., & Hart-Cooper, W., Chan, K.L., Cheng, L.W., Orts, W.J., & Johnson, K. (2016). Antifungal efficacy of octylgallate and 4-isopropyl-3-methylphenol for control of *Aspergillus. Microbial Discovery, 4,* 2.

The search is on for new cosmetic preservatives. *Chemical & Engineering News, 96*(39), (September 30, 2018), 30. Available at: https://cen.acs.org/business/consumer-products/search-new-cosmeticpreservatives/96/i39.

Further reading

Aobiome. https://www.aobiome.com/.

ARS. https://www.ars.usda.gov/pacific-west-area/albany-ca/wrrc/.

BCGC. https://bcgc.berkeley.edu.

Beauty Counter. https://www.beautycounter.com.

Bernauer, U., Bodin, L., Celleno, L., Chaudhry, Q. M., Coenraads, P. J., Dusinska, M., & Wijnhoven, S. (2016). *SCCS OPINION ON phenoxyethanol*. European Union. Available at: https://ec.europa.eu/health/scientific_committees/consumer_safety/docs/sccs_o_195.pdf.

Biomimicry. https://biomimicry.net/.

CEN. https://cen.acs.org/articles/94/i16/Preservative-leavecosmetics-banned-EU.html.

CEN. https://cen.acs.org/business/consumer-products/Preservatives-challenge-awards-cash-prizes/96/web/2018/08.

Chem view. https://chemview.epa.gov/chemview/proxy?filename=8EHQ-18-21337_06.22.2018_Combined.pdf.

CIR Safety. https://www.cir-safety.org.

Clean Production. https://www.cleanproduction.org/programs/greenscreen.

De Groot, A., & Herxheimer, A. (1989). Isothiazolinone preservative: Cause of a continuing epidemic of cosmetic dermatitis. *The Lancet, 333*(8633), 314—316.

ECETOC. https://www.ecetoc.org/links/cpsq-consumer-products-safety-quality-unit-formerly-known-as-european-chemicals-bureau-ecb-part-of-the-institute-for-health-and-consumer-protection-ihcp-joint-research-centre/.

EPA. https://www.epa.gov/pesticide-contacts/contacts-office-pesticide-programs-antimicrobials-division.

Europa. https://ec.europa.eu/health/ph_risk/committees/04_sccp/docs/sccp_s_04.pdf.

EWG. https://www.ewg.org/skindeep/ingredients/704811-PHENOXYETHANOL/.

Facebook. https://www.facebook.com/307128722674171/posts/an-update-from-seventh-generation-thanks-for-this-lisa-yu/1523476907706007/.

FDA. https://www.fda.gov/cosmetics/cosmetics-laws-regulations/fda-authority-over-cosmetics-how-cosmetics-are-not-fda-approved-are-fda-regulated.

FDA. https://www.fda.gov/regulatory-information/laws-enforced-fda/federal-food-drug-and-cosmetic-act-fdc-act.

Green Chemistry and Commerce. https://greenchemistryandcommerce.org/newsletter/2018/august/preservatives-project-winners.

Green Chemistry and Commerce. https://greenchemistryandcommerce.org/projects/innovation/preservatives-project.

Kimura, K., Kameda, Y., Yamamoto, H., Nakada, N., Tamura, I., Miyazaki, M., & Masunaga, S. (2014). Occurrence of preservatives and antimicrobials in Japanese rivers. *Chemosphere, 107*, 393—399.

Marti, E., Jofre, J., & Balcazar, J. L. (2013). Prevalence of antibiotic resistance genes and bacterial community composition in a river influenced by a wastewater treatment plant. *PloS One, 8*(10), e78906.

Safecosmetics. http://www.safecosmetics.org/get-the-facts/chemicals-of-concern/phenoxyethanol/.

Safermade. https://www.safermade.net/.

Seventh Generation. https://www.seventhgeneration.com/home.

Statista. https://www.statista.com/outlook/cmo/beauty-personal-care/united-states.

Toxicology. https://toxicology.vetmed.ufl.edu/.

Toxnet. https://toxnet.nlm.nih.gov/cgi-bin/sis/search/a?dbs+hsdb:@term+@DOCNO+8200).

USPBPEP. http://www.uspbpep.com/usp29/v29240/usp29nf24s0_c51.html.

Chapter 4

Green Chemistry case study on textiles: replacing the toxic cross-linkers formaldehyde, fluorocarbons, and di-isocyanates for wrinkle resistance and water repellency

What was the challenge

This chapter is about work to offer alternatives to some of the more toxic compounds used in manufacturing in the textiles and clothing sector. Some of the worst offenders are used to treat fabric or clothing for wrinkle resistance (WR), and for durable water repellency (DWR). This chapter details the results of an investigation of alternatives to formaldehyde, fluorocarbons, and di-isocyanates used in these treatments.

The Berkeley Center for Green Chemistry (BCGC) and its affiliates have collaborated with Levi Strauss and Company (Levis) on a range of projects from 2013 to 2018. BCGC's sustainable textiles research has proceeded through multiple projects: a fall, 2013, Greener Solutions research partnership with Levis; student-led textile-based research projects in 2014 and 2015; case studies on hazardous chemicals in the textile industry published in 2016 and 2018; and a conference held in 2017.

Initially, Levi's presented two student teams in the Greener Solutions course with two related challenges: first, to identify safer textile finishing compounds for wrinkle resistance (WR) that would allow 100% cotton pants to maintain a smooth visual appearance all day and after every wash, and secondly, to propose a durable water-repellency (DWR) fabric finish that didn't use fluorocarbons, formaldehyde, or di-isocyanates.

Green Chemistry in Practice. https://doi.org/10.1016/B978-0-12-819674-8.00005-9
67

Ultimately both technical issues stemmed from the molecular makeup of cotton fibers in the fabric used to make clothing. Cotton fabric is quite absorbent of water due to the hydrogen atoms contained in its cellulose-based fibers. These positively charged atoms readily attach themselves to the negatively charged oxygen atoms in water, a well-known polar molecule. This distorts the chains of cellulose and warps the fabric. Ironing a shirt to remove wrinkles is applying heat and pressure to reexcite and scramble these atoms back into their original alignment. Similarly, this high absorbency means that water is soaked into the fabric rather than being shed.

Typical treatments for both technical issues have involved interfering with the bonding that occurs naturally between water and cotton. For example, the use of fluorocarbons to treat cotton for durable water repellency (DWR) is effective because of the strength of the covalent bonds between fluorine and other atoms like hydrogen, thus interfering with the weaker hydrogen bonding occurring with water. The strong covalent bonds of fluorine also make substances made from it very hard to break up, however, and thus persistent and potentially harmful to the environment.

In summary, each of the two teams would have to propose alternatives to current WR and DWR chemicals that were less hazardous, of comparable technical and durability performance, and of minimal disruption to the textile application process, cost, and consumer experience of the final product.

Why this project was important

The garment and textile industries are large and growing: the global textile market, worth $1.254 trillion in 2015, had been forecasted to top $1.6 trillion by 2020, due largely to rising consumer demand in China and India. Performance features like durable water repellency (DWR) and wrinkle resistance (WR) are important added values for consumers and are increasingly expected as performance features The current standard chemical approaches to impart fabric with these qualities, however, make use of crosslinking chemicals that pose significant hazards to human health.

Consumers are becoming increasingly aware of safety concerns and are shaping their buying habits around transparency of ingredients, safer substances, and to a lesser extent, concerns about the environment and sustainability. Brand name companies and retailers have responded, particularly in the textile and garment sectors, where companies (which typically share upstream production facilities) have banded together to address common concerns. Industry organizations like the American Apparel and Footwear Association (AAFA), the Sustainable Apparel Coalition (SAC), Zero Discharge of Hazardous Chemicals (ZDHC), AFIRM, and Cradle to Cradle (C2C) have formed to meet the need for resource sharing and collective action.

Levi's is a member of AFIRM, AAFA, ZDHC, and SAC, and is well known for developing its own in-house innovations in chemicals management. The Zero Discharge of Hazardous Chemicals (ZDHC) consortium, for example, was founded in 2011 and includes dozens of other signatory brands, value chain affiliates, and associate organizations. They share the goal of reducing and eventually eliminating hazardous chemical emissions from the textile, leather, and footwear industries.

Recognizing the potential for environmental impact and the value that consumers place on safer chemicals in clothing, Levis joined the ZDHC apparel industry collaboration in 2012, and sought collaboration with BCGC for additional insight into achieving their ZDHC goals. Levis corporate management had assessed the advantages of managing hazard (using safer chemicals to start with) over managing risk (overseeing procedures and practices) within their complex, international supply chain.

The initial challenge to the Greener Solutions course included finding alternatives to the below chemicals of concern, just three in the list of over 600 chemicals used typically in the apparel industry that are considered harmful.

Formaldehyde is used in the production of dyes and pigments, for dye fixing and for imparting wrinkle-free characteristics to cotton and polyester blend fabric during wear and laundering, and for binders and fixers for printing on fabric. It is readily absorbed in the respiratory tract or through the skin. It is carcinogenic (Group 1 carcinogen, IARC), causes dermatitis, and is a respiratory and skin sensitizer. Formaldehyde is on the ZDHC Manufacturing Restricted Substances (MRS) Candidate List. At Levis formaldehyde was used in permanent press fabric finishing in their Docker line of men's clothing. Typically it was used as a base for DMDHEU (dimethylol dihydroxyethyleneurea), and represented primarily an occupational risk to workers.

Flurocarbons are used in oil and water repellency (OWR) treatments and many can be persistent, bioaccumulative, and toxic (PBT) and have been linked to brain tumors. PFOAs Perfluorooctanoic acid (PFOA) and related substances and PFOSs, Perfluorooctane sulfonate (PFOS), and related substances, are on the ZDHC's Manufacturing Restricted Substances List (MRSL).

Di-isocyanates are used in polyurethane coatings and they can be found in synthetic leather and layered fabrics production. Isocyanates include compounds classified as potential human carcinogens (TDI Group 2B) and are known to cause cancer in animals. The main effects of hazardous exposures are occupational asthma and other lung problems, as well as irritation of the eyes, nose, throat, and skin. Often found in vapor or aerosol form during production, these compounds are of most concern to exposed workers. At Levis, di-isocyanates were used to impart a water-repellent finish in so-called commuter jeans.

Who were involved

Levis first approached the BCGC executive director (now former) and current board member Dr. Martin Mulvihill in early 2013, seeking his help to translate Anastas and Warner's 12 Principles of Green Chemistry into a paradigm more readily applicable to apparel production. Mutual satisfaction with the project led to Levis participating as a research partner in the BCGC flagship course, Greener Solutions. During the fall, 2013, Greener Solutions course, Levis challenged students to identify chemical solutions for a water-repellant finish for denim or a permanent press finish for cotton pants without toxic cross-linking compounds (formaldehyde, di-isocyanates, fluorocarbons).

During this phase, the course was led by Mulvihill and physician Dr. Meg Schwarzman of the UC Berkeley School of Public Health, with lecturing assistance from attorney Claudia Polsky of UC Berkeley's School of Law. Amanda Cattermole represented Levis. Dr. Mark Dorfman, of Biomimicry 3.8, advised on bio-inspired examples of chemical cross-linking.

Two student teams were formed; Green and Blue. Both teams investigated cross-linking innovation in cotton fabric for wrinkle resistance (WR) and durable water repellency (DWR), and both relied heavily on molecular bonding models from nature. Team Green members were Joe Charbonnet, Jennifer Lawrence, Leah Rubin, and Sara Tepfer, Team Blue also looked into biomimetic alternatives to fabric finishing. Team members were Michel Dedeo, Lila Lino-Rubenstein, Antony Kim, Kathryn Strobel, Sara Tischhauser, and Katherine Tsen.

What was the solution

Team Green

Team Green proposed a two-step approach to accomplishing both WR and DWR: starting with a binding step that increased the cross-linking ability of the cellulose in the fibers of cotton, followed by cross-linking treatment combinations inspired by nature. Five combinations were proposed and assessed for technical and EHS performance.

Team Green first identified 12 examples of bonding in nature with the help of Dr. Dorfman and characterized them into four themes: covalent bonds, noncovalent bonds, metal-containing bonds, and structural bonds, and noted that it is rare to see any of these types of bonds work in isolation. They focused attention on three of the four themes, discarding metal-containing bonds as outside the scope of their work (Table 4.1).

Their investigative question was whether a combination of bonding or crosslinking types could be employed in a two-step process to effectively replace the traditional one-step covalent bonding currently being used for WR and DWR. Could something be first bonded to the cellulose in the yarns in the first step that would then be bonded to another material? This bonding would either link the yarns to each other (WR) or link the yarns to a water repellent (DWR) (Fig. 4.1).

While the two-step approach afforded some freedom in material choice and timing of application, it also presented the possibility of interference during the numerous intermediate steps in typical garment treatment: high temperatures, strong bases, and strongly oxidizing bleaches might be problematic. The team mapped the typical production process and identified intervention points within this process system where innovation might occur (Fig. 4.2).

TABLE 4.1 Examples of crosslinking within each of the four bonding themes.

Bond Type	Strategy in Nature	Place of Use
Covalent Bonds	Disulfide bonds	- Chinese soft shelled turtle
		- Human joint
		- Malaysian tree frog
		- Pearl Oyster
	Imine bonds	- Slug
	Peroxide-mediated bonds	- Flax stem
Non-Covalent Bonds	Hydrogen bonds	- Chinese soft shelled turtle
		- Chiton tooth
		- Deep sea sponge
		- Flax stem
		- Malaysian tree frog
		- Woody plants
Metal-Containing Bonds	Coordination complexes	- Blue sea mussel
		- Malaysian tree frog
	Direct metal crosslinking	- Slug
	Metal ion interfaces	- Chiton tooth
		- Human joint
Structural Bonds	Anti-parallel sheets	- Chinese soft shelled turtle
	Checkerboard reinforcement bonds	- Deep sea sponge
	Flexible coils	- Blue sea mussel
		- Vineyard snail
	Large, branched structures	- Flax stem
	Micelles	- Chinese soft shelled turtle

(2013). New approaches in cotton crosslinking green team final report to Levi Strauss & co. Unpublished report.

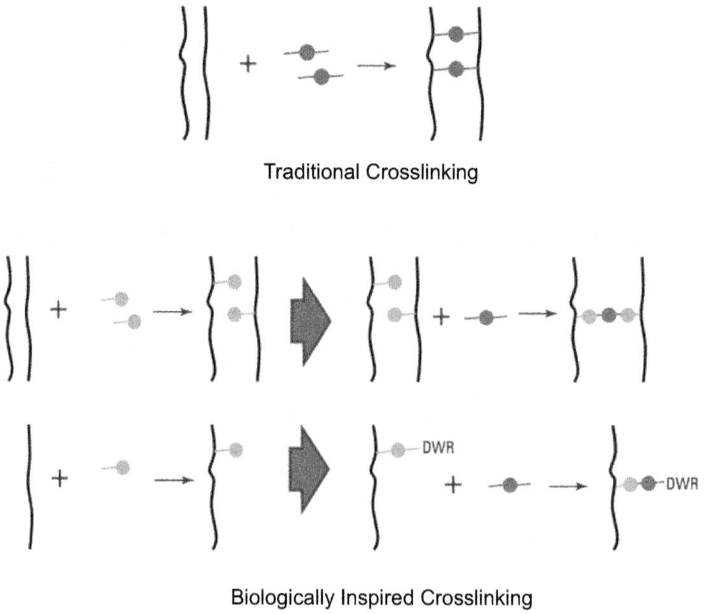

Traditional Crosslinking

Biologically Inspired Crosslinking

FIGURE 4.1   One-step crosslinking compared to potential two-step processes. *Adapted from Charbonnet, J., Lawrence, J., Rubin, L., & Tepfer, S. (2013). New approaches in cotton crosslinking green team final report to Levi Strauss & co. Unpublished report available at http://bcgc.berkeley. edu/greener-solutions-2013/.*

	STEP	SUBPROCESS OVERVIEW	CHEMICAL INPUTS
YARN FORMATION	RAW COTTON		
	FIBER PREPARATION	WAX REMOVAL - hydrophobic coating is removed to allow dyes to penetrate into cellulose	
	SPINNING		

	STEP	SUBPROCESS OVERVIEW	CHEMICAL INPUTS
FABRIC FORMATION	WARPING	Lengthwise yarns are held in tension and weft yarns are later woven through during weaving	
	SIZING	Formulations added to warp yarns to reduce friction and fraying; adds at least 20% by weight.	polyvinyl alcohol polyacrylic acid carboxymethyl cellulose
	WEAVING		For Levi's - indigo dye applied prior to weaving

KNITTING

	STEP	SUBPROCESS OVERVIEW	CHEMICAL INPUTS
WET PROCESSING	PREPARATION	May include: desizing, scouring, bleaching (for pale fabrics), and/or mercerizing	
	DYEING / PRINTING		For Levi's - indigo dye applied prior to weaving
	FINISHING	May include DWR or permanent press applications; traditionally spray or dip process, though more recently CVD. LS&Co. uses a dip process.	DMDHEU; Diisocyanates; perfluorinated acids; fluoropolymer; paraffin-based DWRs

STEP	SUBPROCESS OVERVIEW	CHEMICAL INPUTS
DESIZING	Removal of starches and other water soluble compounds in sizing formulations	amylase other enzymes
SCOURING	Remove fats, waxes, tannins, pectins, proteins and any dirt or husk residues	boiling conc. NaOH 1 hr.
BLEACHING	Remove naturally occurring pigments	100° C H_2O_2
MERCERIZING	Swells fibers in diameter and shrinks them longitudinally, making them more receptive to dyes	25-30% caustic soda

	STEP	SUBPROCESS OVERVIEW	CHEMICAL INPUTS
FINISHING	CUTTING		
	SEWING		For Levi's - DWR properties imparted here (garment form)
	CURING		
	FINISHED GOODS		Lifecycle performance - 10-30 washes, depending on line

FIGURE 4.2   Textile manufacturing process. *Adapted from Charbonnet, J., Lawrence, J., Rubin, L., & Tepfer, S. (2013). New approaches in cotton crosslinking. Unpublished report available at: https://bcgc.berkeley.edu/greener-solutions-2013/.*

Step one- cellulose binding

The team believed that the crosslinking capability of the cellulose base could be improved. Hydroxyls are cellulose's most available functional group but exhibit relatively low reactivity. They could be replaced or augmented with other more reactive molecules and receptive sites for cross-linking in step two. They examined strategies for improving this capability including noncovalent, covalent, and structural bonds.

For noncovalent bonding techniques they outlined three different ways to boost the cross-linking capacity of the cellulose: the current practice of coating fibers with sizing, weaving poly blends into the fabric and forming new material in situ on the fabric.

They also offered alternative approaches using covalent bonding. For example, for covalent bonding to cellulose, they researched well-documented polycarboxylic acids, molecules found in many food substances such as citric and malic acid. When heated, they form cyclic anhydride intermediates that react with the hydroxyl groups on cellulose to form ester linkages. Many of these polycarboxylic acid techniques, although promising and initially bio-based, had drawbacks that included uncompetitive cost, yellowing, and loss of tensile strength in the yarn. Similarly, modifiers and catalysts employed in these techniques had health or environmental costs. For example, the industry standard catalyst for polycarboxylic acids is sodium hypophosphite (SHP). SHP can degrade the cellulose fibers, cause dye color change and algae growth, and eutrophication of waterways when released in waste streams.

The team proposed using a difunctional carboxylic acid, such as succinic acid, which can form a single linkage with cellulose. The acid would be applied and linked to the fabric in the finishing step. At this point, an extra carboxylic acid group would be available for further reactivity. Then common bioconjugation reagents would be employed to react with this carboxylic acid in situ to attach a thiol group. As an alternative, they proposed using a diamine to crosslink with the carboxylic acid groups, either through reductive amination or amidation. When combined with the two-step process this was a novel and untried approach, but the team believed it could open new possibilities for investigation (Fig. 4.3).

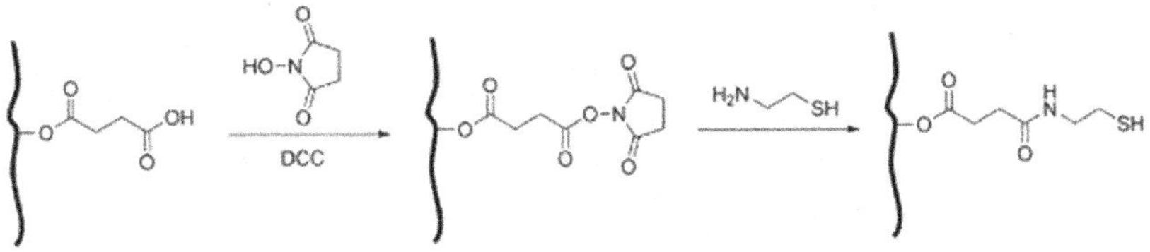

FIGURE 4.3 In situ thiolation of poly(carboxylic acid. *From Charbonnet, J., Lawrence, J., Rubin, L., & Tepfer, S. (2013).* New approaches in cotton crosslinking. *Unpublished report available at: https://bcgc.berkeley.edu/greener-solutions-2013/.*

Step two- cross-linking of either yarns for WR or hydrophobic substances for DWR

With the base cellulosic fibers "primed" for better cross-linking, the team proposed two possible mechanisms for wrinkle resistance and durable water repellency. Both of these mechanisms have models in nature.

Disulfide bonds are commonly found in the formation of secondary and tertiary protein structures in biological systems. They are formed by the oxidative reaction of two systeine residues. This formation requires an oxidizing agent, typically oxygen, a redox-active cofactor, or another thiol. The team proposed that thiol groups be attached to the cellulose through a thiolated polymer coating or by reaction with a poly(carboxylic acid) in the finishing step of the production process. They offered a range of oxidizing agents that could be used.

Imine and amine bonds are also found in the natural world. For example, imine bonds are formed reversibly in the slug, *Arion subfuscus*, to harden the mucus secreted by the animal. This is a dehydration reaction and is easily hydrolyzed back to the original amine and carbonyl groups. While an advantage to the slug, this would be problematic for textile production, so the team recommended an additional step, a reductive amination, using a catalyst and reducing agent that would turn the intermediate imine into a more stable amine.

The team proposed crosslinking by using a diamine crosslinker with a carbonyl-containing cellulose treatment; potentially polyethyelene terephthalate (PET) woven with the cotton, a polymeric fabric coating such as poly(lactic acid) (PLA), or a poly(carboxylic acid) (PCA) that could bond covalently to cellulose. The diamine would be applied during the finishing step, and crosslinking would be accomplished with the addition of the reducing agent and catalyst.

Technical performance

One of the tenets of the Greener Solutions approach is to assess both the technical and EHS performance of alternatives from the outset and to be clear about the relative weighting of the criteria. Teams are encouraged to develop and defend their own assessment criteria and scoring frameworks.

Team Green chose eight technical criteria evenly divided between general indicators of innovation and feasibility within current production methods, as well as more specific criteria for success in crosslinking for WR and DWR. They then assessed individual components within the two-step system and combinations of materials and techniques that had been identified (Table 4.2 and Figure 4.4). This assessment showed that imine bonding with PET to be the most promising from a technical perspective.

EHS scorecard

For environmental health and safety criteria, Team Green employed a modified GreenScreen Chemical Hazard Assessment Criteria (Table 4.3). Levis was

TABLE 4.2 Eight technical evaluation factors.

	ADDITIONAL RESEARCH NEEDED	MAJOR HURDLES ANTICIPATED	MINOR HURDLES ANTICIPATED	OPTIMIZATION ONLY
INNOVATION	CHEMICAL SUPPLY	SPECIAL MANUFACTURE	LIMITED AVAILABILITY	WIDE AVAILABILITY
DISRUPTION OF INFRASTRUCTURE	FABRIC APPLICATION	NEW PROCESS	MODIFY EXISTING PROCESS	USES EXISTING PROCESS
	CROSSLINKING STEP	NEW EQUIPMENT	NEW CHEMICALS, SOLVENTS	HEAT OR AIR CURED
	CONTROLLABLE CROSSLINKING	TOO REACTIVE OR UNREACTIVE	SPECIAL CONDITIONS OR EXTRA CHEMICALS	ADD CATALYST, REAGENT, HEAT
ROBUSTNESS	RESILIENCE DURING MANUFACTURING	LIKELY PROBLEMS	POSSIBLE PROBLEMS	NO FORESEEABLE PROBLEMS
	RESILIENCE DURING CONSUMER USE	LIKELY PROBLEMS	POSSIBLE PROBLEMS	NO FORESEEABLE PROBLEMS
SIDE EFFECTS	EFFECTS ON FABRIC	REQUIRES PROBLEM CHEMICALS	POSSIBLE NEED FOR PROBLEM CHEMICALS	NO FORESEEABLE PROBLEMS

From Charbonnet, J., Lawrence, J., Rubin, L., & Tepfer, S. (2013). *New approaches in cotton crosslinking.* Unpublished report available at: https://bcgc.berkeley.edu/greener-solutions-2013/.

TABLE 4.3 Health evaluation data for the proposed polymer thiolation strategy.

CHEMICAL	FUNCTION	GROUP I HUMAN					GROUP II + II* HUMAN							E TOX		FATE		PHYS	
		C	M	R	D	E	AT	ST sg|rep	N sg|rep	SnS	SnSR	IrS	IrE	AA	CA	P	B	Rx	F
POLY(METHACRYLIC ACID)	POLYMER																		
POLY(ACRYLIC ACID)	POLYMER																		
CHITOSAN	POLYMER																		
CARBOXY METHYL CELL...	POLYMER																		
L-CYSTEINE	THIOLATING AGENT																		
CYSTEAMINE HCl	THIOLATING AGENT																		
THIOGLYCOLIC ACID	THIOLATING AGENT																		
N-HYDROXYSUCCIN...	CATALYST																		
N-(3-DIMETHYLAMINO...	CATALYST																		
DITHIOTHREITOL	REDUCTANT																		
SODIUM BOROHYDRIDE	REDUCTANT																		
POLY(L-CYSTEINE)	THIOLATED POLYMER																		

From Charbonnet, J., Lawrence, J., Rubin, L., & Tepfer, S. (2013). *New approaches in cotton crosslinking*. Unpublished report available at: https://bcgc.berkeley.edu/greener-solutions-2013/.

currently using this hazard-based system, which had been developed from the international Globally Harmonized System (GHS). This is a benchmark approach that evaluates individual chemicals based on data gathered from the literature, modeled data, data for suitable chemical analogs, and hazard lists published by governmental agencies, academic researchers, and nongovernmental agencies. Each chemical and its potential transformation products are evaluated for 18 different endpoints sorted into five hazard groups. Each endpoint is assigned a hazard classification, ranging from Very High to Very Low, and confidence level. These hazard classifications were aggregated into a benchmark score based on the GreenScreen's weighting system (Figs. 4.4 and 4.5).

Five combined strategies were assessed, including baseline chemicals currently being used in the wet production process, like formaldehyde and diisocyanates; disulfide bond crosslinking; amine/imine bond crosslinking; (poly) carboxylic acid process: polymer thiolation; polymer coating process. Despite the high number of data gaps in toxicological endpoint scoring, the team could recommend imine/amine crosslinking with PET as the highest performing from a technical standpoint and the thiolated polymer coating as the safest.

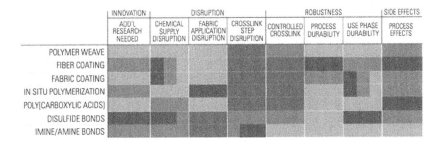

Figure 14: Technical evaluation of each strategy.

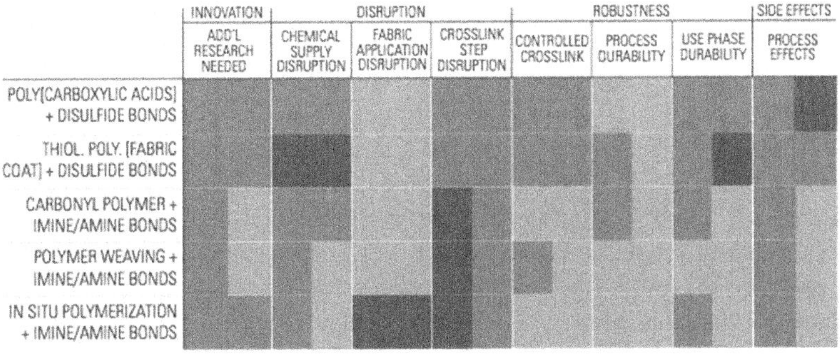

FIGURE 4.4 Technical evaluation of combined strategies. *From Charbonnet, J., Lawrence, J., Rubin, L., & Tepfer, S. (2013).* New approaches in cotton crosslinking. *Unpublished report available at: https://bcgc.berkeley.edu/greener-solutions-2013/.*

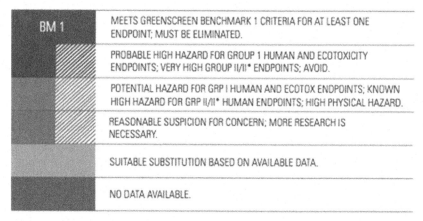

FIGURE 4.5   Health evaluation weighting scheme. *From Charbonnet, J., Lawrence, J., Rubin, L., & Tepfer, S. (2013).* New approaches in cotton crosslinking. *Unpublished report available at: https://bcgc.berkeley.edu/greener-solutions-2013/.*

Team Green's findings comprised a two-step approach to two technical performance issues, WR and DWR by first linking modifiers to cellulose fibers and then crosslinking fibers for performance, be it for water repellency or resistance to fiber misalignment. In a novel approach, the team outlined a dozen mechanisms found in nature, and, importantly, linked these mechanisms to textile processing mechanisms. They also developed, de novo, a rational rating system for judging textile technical performance in their alternatives assessment, and used a modified GreenScreen benchmark ranking system to assess EHS performance. No clear "winner" emerged, with nuanced tradeoffs for each candidate cited, but the team was able to model a process for textile manufacturers to perform a similar alternatives assessment, and point to encouraging material/process paths that Levis could pursue further.

Team Blue

Team Blue focused its attention on the combined use of enzymes and novel substrates gleaned from crosslinking processes found in nature. Their aim also was to identify safer alternatives to formaldehyde resins and di-isocyanates used for WR and DWR. After establishing a technical and EHS assessment framework, they compared four alternative combinations with the baseline "bad actors", and were able to recommend two promising combinations and advise on pathways to overcome associated tradeoffs. They established reducing hazards as their top priority, followed by achieving parity with existing technical performance and durability as their second priority, followed by the application process, cost, and consumer expectations (Tables 4.4 and 4.5).

The four alternatives found in nature were: the polysaccharide monooxygenase (PMO) enzyme; the Lacasse enzyme associated with lignin;

TABLE 4.4 Evaluation metrics for technical feasibility.

	Crosslinking Ability			Durability			Application and Curing				Cost		Consumer Expectations	
	With Cellulose	With Itself (DP)	Add Func. Groups (DWR)	Stable Through Multiple Washes	Fabric Strength	Withstands foods, sun, etc.	Controllably Cured	Time of Curing	Chemical Stability & Water Solubility	Existing Process Machinery	Per kg Raw Material	Per yard fabric	Color	Consumer Trends
Green	Multitude of strong interactions	Multitude of strong interactions	Multitude of strong interactions	No negative interactions with detergents and bonds are	No acid or other known weakening treatments	No issues	Very controllable	<1 hour	No Refrigeration or other special storage, water soluble	Existing machinery can be used	Same price or cheaper than existing tech	Same price or cheaper than existing tech	No color change	Huge marketing plus!
Red	No or unknown interactions	No or unknown interactions	No or unknown interactions	laundry detergent has the potential to undo the crosslinking	weakens fabric significantly	Not stable to something that would contact clothing during normal wear	no control at all (as soon as its on the fabric it will react)	Over 12 hours	requires special storage and not water soluble	New machinery needed	More than double existing tech	More than double existing tech	Large color change	They would hate it

From Dedeo, M., Iino-Rubenstein, L., Kim, A., Strobel, K., Tischhaouser, S., & Tsen, K. (2013). Identifying greener solutions: Biomimetic chemical alternatives for fabric finishing at Levi Strauss & Co. Unpublished report available at: https://bcgc.berkeley.edu/greener-solutions-2013/.

TABLE 4.5 Evaluation metrics for human health & environmental hazard endpoints.

		Evaluation Metrics	
	High □	Moderate ○	Low □
Human Health Group I — Carcinogenicity Mutagenicity Reproductive Developmental Toxicity	Known or presumed for any route of exposure; authoritative lists, strong weight of evidence (human)	Suspected for any route of exposure; limited or marginal evidence (animal)	Adequate data, or clear evidence of no effect
Endocrine	Evidence of endocrine activity and related human health effect	Evidence of endocrine activity	Adequate data available; negative studies
Acute Toxicity	GHS category 1,2,3; any route of exposure;	GHS category 4; any route of exposure	GHS Category 5, adequate data, negative studies, or GHS not classified
AT Oral LD50 (mg/kg)	0-300	>300 – 2000	> 2000
Human Health Group II — Systemic Toxicity Organ Effects	GHS category 1,2 single exposure for any route of exposure	GHS category 3 for single exposure any route of exposure	Adequate data available, negative studies, GHS not classified
Neurotoxicity	GHS category 1,2; single exposure any route	GHS category 3; single exposure any route	Adequate data, negative studies; not classified
Sensitization	High frequency of occurrence	Low to moderate frequency of occurrence	Adequate data, negative studies, not classified
Skin Irritant	GHS category 1,2 (Corrosive/irritating)	GHS category 3 (Mild irritant)	Adequate data, negative studies, not classified
Eye Irritant	GHS category 1,2 (irreversible/irritating)	GHS category 3 (Mild irritant)	Adequate data, negative studies, not classified

Source: Adapted from GreenScreen for Safer Chemicals (24)

		Evaluation Metrics	
	High □	Moderate ○	Low □
Environmental Toxicity & Fate — Persistence — Soil ($t_{1/2}$ life in days)	60 to >180	16 to 60	<16 or rapid degradability
Water ($t_{1/2}$ life in days)	40 to >60	16 to 40	< 16 or rapid degradability
Air ($t_{1/2}$ life in days)	>2 to 5	-	<2
Long-Range Environmental Transport	Evidence	Suggestive evidence	--
Bioaccumulation — BAF (Bioaccumulation factor)	1000 to >5000	>500 to 1000	0 to 500
BCF (Bioconcentration factor)	1000 to >5000	>500 to 1000	0 to 500
Log K_{ow} (Log octanol-water partition coefficient)	4.5 to > 5.0	>4.0 to 4.5	≤4
Monitoring Data	Evidence	Suggestive Evidence	--
Aquatic Toxicity (chronic/acute)	GHS category 1,2	GHS category 3	Sufficient data available and not classified
Physical Hazards — Reactivity (Rx)	GHS unstable, category 1,2	GHS category 3; any route of exposure	Adequate data, and GHS not classified
Flammability	GHS category 1,2	GHS category 3,4	Adequate data, and GHS not classified

From Dedeo, M., Iino-Rubenstein, L., Kim, A., Strobel, K., Tischhauser, S., & Tsen, K. (2013). *Identifying greener solutions: Biomimetic chemical alternatives for fabric finishing at Levi Strauss & Co.* Unpublished report available at: https://bcgc.berkeley.edu/greener-solutions-2013/.

dopamine associated with the formation of mussel byssal; potassium permanganate, associated with slug mucus. A summary of these candidates is shown in Tables 4.6 and 4.7.

These candidates were compared to the baseline, "bad actor" formaldehyde resins and di-isocyanates for technical performance across a range of evaluation metrics: crosslinking ability, durability, application and curing, cost, and consumer expectations. A summary of the technical feasibility and EHS scorecards is shown in Table 4.8.

From a strictly technical performance standpoint, the PMO enzyme ranked lowest for consideration based on lack of commercial availability. The remaining three alternatives, the Lacasse enzyme (lignin inspired), dopamine (mussel inspired), and potassium permanganate (slug inspired) still possessed drawbacks, chiefly discoloration of the fabric.

EHS scorecard

Team Blue also used the GreenScreen for Safer Chemicals Comparative Hazard Assessment Criteria as a model to build a simplified version, assessing 18 endpoints, collecting additional exposure data, and summarizing scores in a three-tiered color-coded benchmark ranking system.

The team performed a four-tiered sequence of research for toxicological data, starting with authoritative lists compiled by Pharos Chemical and Material Library of the Healthy Building Network, as well as using PubMed, Google Scholar, and Web of Science. This was followed by a review of resources such as the Hazardous Substances Data Bank (HSDB), EPA Integrated Risk Information System (IRIS), and Agency for Toxic Substances Disease Registry (ATSDR) Toxic Substances Portal. The team viewed Material Safety Data Sheets (MSDS) and product information from manufacturers as a last resort. In circumstances with no data, they employed some limited modeling to estimate environmental hazard parameters like bioaccumulative potential and persistence.

During this period of discovery, the team observed and documented information about their baseline "bad actors", noting that isocyanates as a class of chemicals are both sensitizers and irritants; toluene di-isocyanate, in particular, is a probable human carcinogen and that Formaldehyde is a Benchmark One chemical, and is listed as a Class 1 carcinogen by the International Agency for Research on Cancer (IARC).

In this testing regime, the enzyme PMO scored the highest, rather than the lowest (as in the technical performance testing), and the group judged it and the mussel-inspired solution (Dopamine crosslinker), to be the safest choices. These kinds of contradictions are not uncommon, and typically Greener Solutions teams are encouraged to enter the second round of feasibility ranking, making informed judgments about the relative difficulty of mitigation necessary to improve a formula or process.

TABLE 4.6 Summary of solutions.

Solution	Summary of chemistry	Proposed chemicals
Polysaccharide Monooxygenase (PMO) Enzyme inspired by cellulase enzymes	Functionalize cellulose and crosslink with substrate	**Enzyme:** Polysaccharide Monooxygenase **Possible primary amine crosslinkers:** ethanolamine, p-phenylenediamine (PPD), o-phenylenediamine (OPD), ethylene diamine, lysine
Laccase Enzyme inspired by Lignin	Crosslink network around cellulose	**Enzyme:** Laccase enzyme, peroxidase enzyme **Possible substrates:** p-coumaric acid, methyl hydroquinone, 2,5-dihydroxybenzoic acid, 4-aminophenol, vanillin
Dopamine inspired by the Sea Mussel	Crosslink network around cellulose	**Crosslinker:** Dopamine **Buffer salt:** Tris HCL **Water repellent chemical:** octodocylomine
Potassium Permanganate inspired by the Slug	Functionalize cellulose and crosslink with substrate	**Required Chemicals:** potassium permanganate, sodium hydroxide **Substrates:** Iron(II) **Recommended Chemicals for Mixing:** acetone, pyridine, dioxane, t-butyl alcohol

From Dedeo, M., Iino-Rubenstein, L., Kim, A., Strobel, K., Tischhaouser, S., & Tsen, K. (2013). *Identifying greener solutions: Biomimetic chemical alternatives for fabric finishing at Levi Strauss & Co.* Unpublished report available at: https://bcgc.berkeley.edu/greener-solutions-2013/.

TABLE 4.7 Technical evaluation of greener crosslinking solutions compared to current chemistry.

			Evaluation Metrics												
Chemical Compound	Durable Press or DWR	Crosslinking Ability			Durability			Application and Curing				Cost		Consumer Expectations	
		With Cellulose	With Itself (DP)	Add Func. Groups (DWR)	Stable Through Multiple Washes	Fabric Strength	Withstands foods, sun, etc.	Controllably Cured	Time of Curing	Chemical Stability & Water Solubility	Existing Process Machinery	Per kg Raw Material	Per yard fabric	Color	Consumer Trends
Existing Solutions															
DMDHEU	DP														○
DMeDHEU	DP		○		○	○								○	
Citric Acid	DP				○	○								○	
BTCA	DP				○										
Other															
Other															
Proposed Solutions															
Polydopamine	DWR	○			○										
Silane modification of cotton & reaction with aldehydes	Ether	○	○	○			○					○			
Laccase oxidation & radical crosslinking	DP	○						○	○	○				○	
PMO enzyme oxidation of cellulose and reaction with amine	DWR		○					○		○					
Permanganate				○	○	?			?						
Rejected Solutions															
4-Arm PEG catechol solution		○	○							○					
Cellulose binding modules															
Polyacrylic acid on cellulose															
Permanganate functionalized cellulose															

From Dedeo, M., Iino-Rubenstein, L., Kim, A., Strobel, K., Tischhaouser, S., & Tsen, K. (2013). Identifying greener solutions: Biomimetic chemical alternatives for fabric finishing at Levi Strauss & Co. Unpublished report available at: https://bcgc.berkeley.edu/greener-solutions-2013/.

TABLE 4.8 Summary of technical feasibility and health and environmental hazards for proposed solutions.

Solution	Water Repellency or Permanent Press?	Technical Benefits	Technical Challenges	Health effects summary
PMO Enzyme inspired by cellulose enzymes	Water repellent	Covalent bond to fabric Variety of chemicals possible	Uncertain level of fabric modification Slow Enzyme is not available commercially	Enzyme- likely safe Amines- generally toxic, lysine is safest because biologically derived
Laccase Enzyme inspired by Lignin	Wrinkle resistant permanent press Water repellent	Enzyme initiated coupling Wide range of substrates can be oxidized and coupled Radical transfer can	Laccases have been used to bleach dyes in the textile industry! Possible color formation	Enzyme- likely safe Possible substrates- generally toxic, vanillin is safest

From Dedeo, M., Iino-Rubenstein, L., Kim, A., Strobel, K., Tischhaouser, S., & Tsen, K. (2013). *Identifying greener solutions: Biomimetic chemical alternatives for fabric finishing at Levi Strauss & Co.* Unpublished report available at: https://bcgc.berkeley.edu/greener-solutions-2013/.

What was innovative about the solutions

All of the chemical crosslinking solutions took inspiration from natural cross-linking chemistry, rather than industrial chemistry, but the researchers used their knowledge of the strengths and limitations of the industrial chemical ecosystem to propose strategies for implementing these solutions at industrial scales. The solutions were not only evaluated for cost and efficacy, but also for sustainability and their effects on human and environmental health. The researchers identified a range of potential solutions for implementation at different steps of the production process, demonstrating a thorough understanding of the challenges associated with textile production, and flexibility in problem-solving. They vetted solutions for feasibility and the most promising solutions and dopamine-based crosslinking, and thiolated cellulose-bound polymers were determined to have the greatest potential to make more sustainable water-resistant, wrinkle-free fabrics. Both of these technologies required additional research and development but presented novel avenues for greener textile solutions.

What was the impact

Once the course concluded Levis continued consulting with the BCGC to investigate two additional aspects of the denim lifecycle: fabric dyeing, and consumer washing habits in 2014 and 2015, respectively. The 2014 fabric dye research collaboration extended research from the Greener Solutions course to explore the use of the laccase enzyme to catalyze crosslinking between the white cotton fabric and common textile additives and dyes. Kathryn Strobel worked with Levis and Dr. Mulvihill to demonstrate the effectiveness of the strategy. Though this research did not yield commercially-viable results, Levis was sufficiently satisfied to invite the BCGC onto a joint research project with the Haas Center for Responsible Business in 2015, in which SAGE fellows Julie Chaves and Jennifer Lawrence conducted extensive literature-based research to answer the question: "is it really necessary for consumers to wash their jeans?"

The question, prompted when the CEO of Levis stated in an interview that he never washed his jeans, was whether there were any microbiological consequences to (wearing but not washing) Levi's products, and if so, what were they? The BCGC researchers Jennifer Lawrence and Julie Chaves found that there were indeed some microorganisms commonly found on cotton that might cause malodor or lead to pathology for the consumer if they were allowed to proliferate, but these species constituted a very small subset of the overall denim microbial population. More importantly, they argued, was controlling the availability of oils and other food sources for bacteria that might otherwise grow on denim. They also noted that many of the adverse effects associated with microbial growth on denim clothes could also be attributed to synthetic chemicals or environmental factors not related to the rate of washing.

In 2016, BCGC board member Dr. Mulvihill partnered with the Center for Responsible Business (CRB) at the Haas School of Business to produce a case study on Levis' Screened Chemistry Program, a hazard-based chemical assessment program, developed in part to help the company reach its ZDHC goals. Dr. Mulvihill partnered with the CRB again in 2017, to host a conference on the Berkeley campus to further discuss the successes of Levis' chemical safety initiatives. In 2018, Dr. Mulvihill coauthored a definitive report on the state of sustainability and Green Chemistry in the fashion industry in partnership with the global initiative Fashion for Good.

BCGC students, faculty, and staff have worked on the chemistry of bio-inspired cotton crosslinking chemistry, sustainable color-fastening, and the microbiology of unwashed jeans. BCGC board members and affiliated researchers have produced sustainability reports on both Levis and the textile industry as a whole, and as recently as June 2018, published a comprehensive analysis of the challenges and opportunities that lie ahead for companies pursuing greener and more sustainable chemistries. These research products span topics across the entire textile product lifecycle for both Levis and for general audiences, providing insights and recommendations on improving the early phases of preparing the cotton for weaving all the way through to durability and aesthetic performance during consumer use.

References

Charbonnet, J., Lawrence, J., Rubin, L., & Tepfer, S. (2013). *New approaches in cotton crosslinking*. Unpublished report available at: https://bcgc.berkeley.edu/greener-solutions-2013/.

Dedeo, M., Iino-Rubenstein, L., Kim, A., Strobel, K., Tischhaouser, S., & Tsen, K. (2013). *Identifying greener solutions: Biomimetic chemical alternatives for fabric finishing at Levi Strauss & Co.* unpublished report available at: https://bcgc.berkeley.edu/greener-solutions-2013/.

Mulvihill, M., & Horotan, A. (2018). *Safer chemistry innovation in the textile and apparel industry*. Unpublished report available at: https://www.safermade.net/reports-and-analysis.

Strand, R., & Mulvihill, M. (2016). *Levi Strauss & Co: Driving adoption of green chemistry*. SAGE Publications Limited.

Further reading

AAFA. https://www.aafaglobal.org.

AFIRM. https://www.afirm-group.com/.

Apparel Coalition. https://apparelcoalition.org.

Business Insider. https://www.businessinsider.com/levis-ceo-dont-wash-your-jeans-2014-7.

C2CCertified. https://www.c2ccertified.org/.

CDC. http://www.cdc.gov/niosh/topics/isocyanates/.

EPA. https://www.epa.gov/pfas/what-are-pfcs-and-how-do-they-relate-and-polyfluoroalkyl-substances-pfass.

IARC, W. (2006). IARC monographs on the evaluation of carcinogenic risks to humans. *Form-aldehyde, 2-Butoxyethanol and 1-Tert-Butoxypropan-2-Ol, 88*, 39–325.

Roadmap. https://www.roadmaptozero.com.

Shenglufashion. https://shenglufashion.com/2017/06/06/market-size-of-the-global-textile-and-app arel-industry-2015-to-2020/.

Chapter 5

Green Chemistry case study on additive manufacturing: finding safer alternatives for stereolithography (SLA) resins: replacing acrylates and methacrylates as cross-linkers

The Berkeley Center for Green Chemistry (BCGC) has been investigating safer alternatives for additive manufacturing (AM) since 2014, and this investigation has comprised six sequential phases with distinctly different collaborative partnerships and slightly different objectives. We discuss each phase as a separate challenge below, but each phase and the discoveries made in it has led to and supported the next, and the outcomes in this case study should be viewed as a sum total of all the efforts.

Phase one: the initial consulting project with Autodesk

In 2014, the software maker Autodesk of San Francisco, California, contracted BCGC to investigate the chemical characteristics of a stereolithography (SLA) resin, PR48, which had been developed by Colorado Polymeric Systems (CPS) of Boulder, Colorado, specifically for Autodesk's demonstration SLA printer, the Ember. The Ember was part of the Spark Investment Fund, a $100M program being promoted by Autodesk to stimulate broad innovation and increase use in the additive manufacturing (AM) market sector and to launch the use of the Spark software product for use in all 3D printers. To that end, the company had designed and marketed the Ember as an open source hardware device, making its specifications freely available to the general public. Autodesk was concerned about the material composition of PR48, however; in particular, its relative level of safety.

What was the challenge

BCGC was asked to characterize the typical range of SLA resin components available, including those in PR48, and to assess for Autodesk the relative

Green Chemistry in Practice. https://doi.org/10.1016/B978-0-12-819674-8.00003-5

89

human and environmental health hazards of these materials. In particular, was it possible for Autodesk to improve the safety performance of its resin? Moreover, the consultant team was asked to investigate and propose biologically inspired material solutions for reformulating or replacing PR48, as part of Autodesk's program of searching for nature-based innovation. It was intended that the results of this study would form the basis for subsequent initiatives at the company, such as a public design competition that would align with their open-source approach to innovation. Of particular concern to management was the cost of toxic waste disposal from unpolymerized resins and the consequences of promoting to the public potentially harmful materials for AM.

In SLA, a liquid resin is cured into a solid set of layers by ultraviolet (UV) light either in a "direct write" method in which a laser traces vectors on the liquid, or in a digital light projector (DLP) method in which a pattern of light is flashed upon each layer. The light sets off the monomers in the liquid to bind together in polymers by first interacting with an initiator molecule that sets off the chain reaction. This photoinitiation of polymerization typically occurs in one of two forms: free radical or cationic polymerization: both form covalent bonds (Fig. 5.1).

A typical resin is comprised of 0.5%−10% of a photoinitiating chemical, < 1% of an optical brightener, and >80% of an organic monomer. The resin will

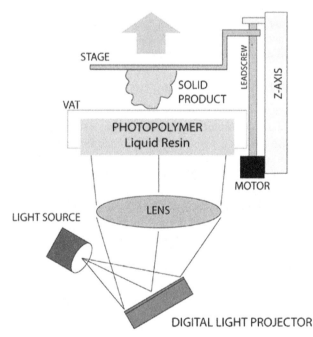

FIGURE 5.1 Typical digital light projector (DLP) arrangement. *From orignal material by author, Thomas A. McKeag.*

also often contain up to 10% of other additives including dyes, colorants, and additives to control shrinkage and other properties of the final plastic.

Why this project was important

Double-digit growth in the AM industry and increased entrepreneurial innovation had followed the expiration of some key patents, but the sector had still not become standardized for either technical or EHS (environmental, health and safety) performance. The AM industry grew 21% in 2017, for example, to top $7 billion in annual sales, the increase coming as 3D printing machines gained traction for both hobbyists and industrial production. Both the volume and the range of materials used in AM were expanding rapidly. The long nascence of and rapid innovations in AM technology and markets presented opportunities for novel technical performance, but concerns were growing about the sustainability aspects of the industry. While mass customization possible with AM as well as decentralized production and just-in-time manufacturing could show significant energy and materials savings, material toxicity and persistence were issues that had yet to be explored, let alone addressed. Moreover, with increasing consumer demand for home and personal 3D printers came increasing potential for exposure to hazardous substances normally mitigated by management and engineering controls within the manufacturing sector. This presented a unique opportunity to attenuate hazard and mitigate environmental costs before industry practices and supply chains had become fully entrenched.

This opportunity included formulations used for SLA resins, many of which are classified as hazardous waste in the State of California. A typical resin formulation may cause skin and eye irritation, be an aquatic and reproductive toxicant and skin sensitizer. Also, a significant percentage of the monomers used in SLA machines are not polymerized; possibly as much as 20%. This means that these materials are loose in the environment at the workplace and more liable to be taken up through inhalation, skin absorption or ingestion.

While there was industry concern for worker and user safety, none of this had coalesced around a set of uniform material safety standards or best practices, let alone an organized campaign for safer substitutions of chemicals of concern in these resins. While the American Society for Testing Materials (ASTM) and the International Organization for Standardization (ISO) had formed committees to prepare standards for material safety associated with the AM sector, these standards were in the preliminary stages of creation, with those for airborne emissions from fused deposition modeling being the furthest advanced.

The search for benign alternatives was aligned with the current industry pursuit of technical innovation and might lead to more efficient performance, cost savings and disruptive technologies, factors that appeared to drive investment in this competitive and free-wheeling market.

Who were involved

The initial Autodesk consulting team comprised the Berkeley Center for Green Chemistry, Biowerks, and the Biomimicry Institute. Dr. Martin Mulvihill, chemist and executive director of BCGC and phD chemistry student Justin Bours represented BCGC; Tom McKeag, principal, represented the bio-design consultancy BioWerks; Beth Rattner, executive director, represented the Biomimicry Institute, an essential liaison to Autodesk.

Autodesk was represented by Dawn Danby, Senior Sustainable Design Program Manager and Susan Gladwin, Sustainability Project Manager. Working meetings were attended by several key technical staff, Materials Scientist Brian Adzima, Global Environmental Health and Safety Manager Julia Cabral, Senior Research Scientist Chris Venter and Senior Experience Designer Shalom Ormsby.

What was the solution

The consultant team assembled a matrix of chemicals used in the typical components of an SLA formula: photoinitiator, reactive monomers and oligomers, UV blocker, dyes, and additives. They then consulted curated lists for toxicological information on common resin bases. In addition, they investigated the state of the industry in four categories of alternative formula base materials: mineral, protein, carbohydrate, and synthetic, and searched for appropriate models for safer substitutions.

They assessed the six acrylates currently being used in the Ember SLA printer, and five alternate monomer types using primary source data. The alternate material types were acrylates, thiols and alkenes, vinyl ethers, epoxides, and oxetanes. They used six criteria for ranking: Sensitization, Acute Toxicity, Carcinogenicity, and Mutagenicity, Endocrine Disruption and Reproductive/Developmental Toxicity, Environmental Toxicity, and Persistence and Bioaccumulation. Each sample was assigned a value of increasing hazard concern from 1 to 4. In many cases, a range of value was assigned and there were data gaps in 18% of the cases (Table 5.1).

In order to assess the hazards associated with each compound, the team searched authoritative regulatory lists, chemical hazard databases, and relevant primary literature for known hazard data. First, researchers identified the relevant chemicals from the primary chemistry literature, patents, and discussions with industry experts. They then relied on sources such as ChemSpider and material safety data sheets (MSDS) to provide an overview and relevant chemical identifiers for each substance in a resin. Next, authoritative sources like governmental, regulatory, or international groups were searched for the most recent hazard or regulatory data on compounds. If a chemical was listed by one of the authoritative sources it was relatively easy to gain access to details regarding relevant hazard categories. If the chemical had not been

TABLE 5.1 Chemical hazard ranking scheme.

Hazard	4	3	2	1
Sensitization	known resp.	suspected resp. / known skin	susp. skin	probably not
Acute Toxicity	very high	high	moderate	low
Carc. / Mutag.	known	suspected	possible	probably not
ED, Repr. / Dev.	known	suspected	possible	probably not
Environ./ Chronic Tox.	very high	high	moderate	low
Persist / Bioacc.	very persist. and bioacc.	very persist.	moderately persist. and bioacc.	low persist. and bioacc.

From: Mulvihill et al. (2015). *Toward proactive materials substitution in the resin formulations for the Autodesk Ember SLA printer: a biomimetic approach.* Unpublished report.

studied by these authoritative bodies, then the team searched the primary literature using PubMed to fill in gaps, and to see if there were any early indicators of concern reported by health experts.

If both authoritative lists and the primary literature were searched thoroughly and no data was found, then certain inferences had to be made in order to estimate hazard. Hazard can be estimated by analyzing the functional groups present in the molecule, understanding how the molecule's chemical class can inform hazard, and making analogies to similar chemicals. Persistence, bioaccumulative ability, and overall toxicity were the three hazard endpoints that could be separately analyzed by these hazard estimation techniques.

The early toxicology research indicated the most salient human and environmental health issues, and this determined the basic performance criteria and investigative approach: bioaccumulation potential needed to be reduced and biodegradation potential needed to be increased while not compromising (or perhaps improving) polymerization rates and speed. These criteria led to research into biomedical applications, particularly tissue engineering and organ repair because of the high value investment, record of research and rigorous standards for health and safety in that sector.

The final, unpublished report, Toward Proactive Materials Substitution in the Resin Formulations for the Autodesk Ember SLA Printer: A Biomimetic Approach (04-23-15), Mulvihill et al., identified several technical challenges in the SLA printing process that had significant impacts on the human and environmental health impacts of the resin. While the acrylate-based PR48 formula was one of the less hazardous materials typically used in SLA resins, it did present cause for caution. The material had been found to be a skin and eye irritant, an aquatic toxicant and, a skin sensitizer. Exposure to uncured resin can cause sensitization causing increasingly dangerous reactions (immune-mediated responses like asthma and rashes) in some users over time and repeated exposure. The BCGC/Biowerks team also noted that 20%−30% of the monomers in PR48 formula were not being polymerized during the printing process and therefore more freely available to be taken up by the body. These unreacted monomers were classified as hazardous waste by the State of California, and represented a liability to the company and the public (Table 5.2).

The team identified 11 biomimetic strategies that could be pursued in subsequent innovation investigations, among them demonstrated concepts such as self-organization, hierarchy across linear scales, composite construction and functionally graded material. In all cases, these concepts were tied to the performance requirements of AM. Table 5.3 is a later summation of the principles outlined in the initial consulting report. Fig. 5.2 is an example of one of these concepts.

Finally, they cited three examples from the biomedical field that had demonstrated success in reducing the bioavailability of the resin components: a PVA blend with vinyl carbonates and carbamates, a Poly(glycerol sebacate

TABLE 5.2 Chemical hazard ranking of Autodesk acrylates.

Autodesk Acrylates	A	B	C	D	E	F
Sensitization	2	3	2	2	2	2
Acute Toxicity	1	2	1	2	2	1
Carc./Mutag.	1	1	DG	1	1	1
ED, Repr. / Dev.	1	1	DG	DG	2	DG
Environ. Tox.	3	3	DG	DG	DG	2
Persist / Bioacc.	2	2	1	1	1	1

From: Mulvihill et al. (2015). *Toward proactive materials substitution in the resin formulations for the Autodesk Ember SLA printer: a biomimetic approach. Unpublished report.*

TABLE 5.3 Bio-inspired design strategies.

Design Principle	Biological Inspiration	Application to SLA
Unity within Diversity: *Minimum parts for maximum diversity*	Organisms are comprised of similar components that combine differently to create vast diversity of life	Components of resin can be combined via SLA process to polymerize in countless formations
Multitasking Monomers: *Relationships matter*	Systems are ubiquitous in nature, and outcomes are influenced by relationships between components	Consider all parts and SLA process involved in the making of the final object
The Optimal Activator	The environment can become the trigger that activates the response	Air, light, temperature and pressure are all present in the SLA process
Taking Advantage of Gradients: *Making Delta do Work*	Interrelationships of environmental conditions reveal energy gradient	Energy associated with a phase change, from liquid resin to solid polymer can be used
Shape is Strength	Forms in nature arise from natural selection towards structural advantage	Shape optimization in SLA saves materials and time; layer adhesion is key.
Self Organization	Self-assembly has structural implications at smaller scales, and often utilizes the polarity of molecules	Photo-initiated polymerization in SLA as a form of self-assembly at the molecular scale
Bottom-Up Construction	Molecular level construction to produce emergent properties, like growth, repair, and adaptation	Elements at the molecular level can be combined to produce emergent properties.
Hierarchy Across Linear Scales	Multiple scale construction allows for strength, durability, and resistance to structural stresses	Resin ingredients polymerized sequentially, possibly through self-assembly
Functionally Graded Materials	Parts are changed across space to respond to stresses or structural challenges	Local composition control (LCC) can save materials and energy in manufacturing
Composite Construction	Balancing strength of mineral matrix and flexibility of protein or polysaccharides to address structural challenges at different scales	Shift in AM from prototyping to higher performance parts
Water is the Universal Medium	Biology is water-based; living things benefit and are limited by water	Water-based materials may allow for extreme recyclability, multiple functionality, common sourcing.

From: Thomas A. McKeag and Cheng et al., 2015. *Autodesk and stereolithography 3D printing: Bio-inspired resins for better human and environmental health*, un-

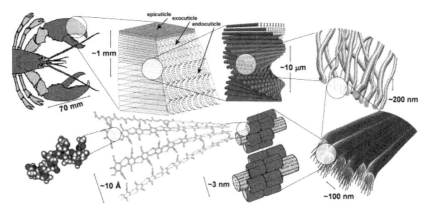

FIGURE 5.2 Example of hierarchical structure across linear scales. *From: Raabe et al. (2005),* Acta Materialia, 53, 4281–4292.

acrylate) resin, and a chitosan-based resin—all of which were predicted to satisfy the remit of improved performance and diminished hazard.

The three examples of polymer research represented the intersection of the biomimetic and green chemistry concepts presented earlier in the final report. The first two examples from the literature were synthetic SLA resins that incorporated design principles that could reduce hazard and increase the biodegradability of the resulting polymer. The third example was a bioplastic based on a naturally occurring polysaccharide. These examples showed how abundant natural materials can be modified and adapted to polymer printing and processing technologies. The team chose two synthetic formulations to illustrate how increasing the molecular weight of the material could reduce cytotoxicity, and adjusting the structure and attachment methods of various sidechains could improve green performance. Their third example was based on chitosan, a derivative of a naturally occurring polysaccharide, chitin, the second most common organic compound in the world. Here, researchers had demonstrated its useability as a bioplastic in casting, molding, extrusion and, to a limited degree, as a photopolymerizable material in micro-SLA.

Example 1: Poly(vinyl alcohol), PVA, as backbone with vinyl carbonates and carbamates

Synthetic monomers for building scaffolds for tissue engineering in *in vitro* medical procedures using micro-SLA.

In this example, an Austrian research team sought to minimize the toxicity of common synthetic SLA monomers, while increasing their biodegradability. The main problem that they had addressed was that too many of the monomers were not being polymerized by the UV light and were subsequently poisoning and killing nearby cells.

This group of researchers worked with various size monomers that incorporated degradable chains connecting the photo-reactive groups. They compared the acrylate, methacrylate, and vinyl versions of these monomers and found that they could increase the molecular weight to reduce the cytotoxicity of these monomers. Larger monomers both reduce the absorption and decrease the biological reactivity of the monomers, further justifying a shift to higher molecular weight monomers in SLA resins. They also found that increasing the molecular weight (size) of the monomers helped to decrease the shrinkage of the resin after photo-polymerization. The biodegradability of these polymers was dependent on both the type of photo-reactive as well as the way the photo-reactive group was attached to the degradable chain. The vinyl and acrylate reactive groups were more degradable then the branched methacrylate reactive groups and esters were the most degradable linkage to the chain.

Example 2: Poly(glycerol sebacate acrylate) PGSA

Synthetic monomers for making tissue repair adhesive in vitro in minimally invasive (medical) procedures (MIP) using micro SLA.

In this example, a team of researchers from MIT and Brigham and Women's Hospital in Boston sought to create a biocompatible adhesive for tissue repair that would stay intact in the aqueous and dynamic environment of surgery before being polymerized by precisely aimed UV light. It would also have to be biodegradable within the body after the temporarily joined tissue had grown new cells to heal the wound.

The team made a new, larger biologically compatible SLA resin using glycerol and sebacic acid which are both biocompatible and are found in most living organisms. The researchers attached photo-reactive acrylates to cross-link the sebacic acid/glycerol polymer when exposed to UV-light. This resin system was used to close tears in high-pressure blood vessels and maintained biocompatibility, durability, and eventual degradability when used to repair cardiac walls in pigs. This is a further example of how high molecular weight (large) monomers can be used to ensure resin safety and performance in biological settings. The viscosity of the resin, however, also increases with molecular weight, which is not always desirable in a benchtop printing device. This was seen as a trade-off that had to be addressed before these environmentally and healthy resins could gain widespread adoption (Fig. 5.3).

Example 3: chitosan

Polysaccharide-based hydrogel for universal use as a bioplastic in heat-assisted injection molding and casting.

In this example, researchers at Harvard's Wyss Institute sought to create an environmentally benign substitute for petroleum-based plastics that could be made from cheap, readily available natural materials.

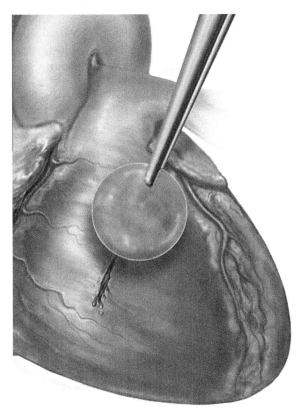

FIGURE 5.3 Surgical glue example. *From: Jeffrey Karp Laboratory, N.D., Massachusetts Institute of Technology.*

The team developed a moldable chitosan bioplastic in a hydrogel formulation. This plastic is processed from deacetylated chitin, a sugar found in arthropod shells and the walls of fungi. Chitin is the second most abundant organic compound in the world, next to cellulose. Unlike cellulose, it can be processed to a useable form without the use of toxic chemicals. The bioplastic has been demonstrated to be useable in both casting and injection molding processes. The researchers used mild conditions including processing in acetic acid (vinegar) to modify the biological polymer so that it would be compatible with current plastic molding technologies and so it would even take up dyes. These dyes can be leached out of the material before recycling by changing the pH.

Hydrogels are networks of polymer chains that are cross-linked and have hydrophilic groups. They can be naturally formed or synthetic. From a phase perspective they are interesting because they can be categorized as neither solid nor liquid, but have properties of both: containing 50%−90% water but not being dissolved by it. They can be formed into many different structures depending on the chemical composition and the methods of cross-linking and initiation. Hydrogels can be cross-linked in three ways: physically, by covalent bonding or ionic bonding. The cross-linking that causes this structural change can be initiated by temperature, pH, and salt concentration.

Hydrogels are often polymers containing carboxylic acid groups. One common polymer used to make hydrogels is sodium polyacrylate. The chemical name for this polymer is poly(sodium propenoate). This coiled molecule with negatively charged oxide side chains can uncoil in the absence of salt as the side chains repel each other. The polymer can absorb about 500 times its weight in water as the positive hydrogen in water molecules bonds to the negative oxygen in the oxide side chains. Despite this high water content, the gel will exist and behave somewhat like a solid because of the three-dimensional cross-linking within the polymer chains.

Natural examples of hydrogels are cartilage, the vitreous humor of the eye, tendons, mucus and blood clots. Natural hydrogel materials being investigated for tissue engineering include agarose, methylcellulose, hyaluron, and chitin.

While this polymer was not demonstrated for use with SLA technology in the referenced article, it had been tested by others as a photo cross-linked hydrogel adhesive for use in neurosurgery It had also been shown by others to be useable in a mechanical pressure actuated deposition type of AM. Here a chitosan-based formula is either premixed or mixed at point of extrusion from a nozzle Additionally, other photo cross-linkable biopolymers, including alginate, hyaluronic acid (HA) and gelatin derivatives, have been printed using SLA. Finally, chitosan has been used for tissue scaffolds in the form of a methacrylamide chitosan hydrogel, although it was not photo cross-linked.

In tissue engineering research poly(ethylene glycol) (PEG), a synthetic hydrogel, has been one of the most widely used formulations for SLA because of its hydrophilicity, biocompatibility, and ability to be chemically tailored. It has to be modified to be biodegradable, however. Chitosan, by contrast, is biodegradable in less than 2 weeks in landfill and has the extra advantage of adding plant growth nutrients to the soil as it degrades.

The future development of processing and material modifications could make chitosan hydrogels more attractive for SLA prototyping applications. The common availability and degradability of this biomaterial, along with its safety to humans and the environment, made this an attractive material to investigate.

What was innovative about the solution

Several aspects of this research and analysis were novel and ground-breaking. To the team's knowledge, no one in the AM sector had ever investigated comprehensively the relative toxicological hazards of a range of typical SLA formulas. This was also the first time that biologically-inspired design strategies were presented as new paradigms for AM processes and materials. Finally, the solutions cited from the field of biomedical research and tissue engineering, used as models because of the high bar for safety and effectiveness, had not been investigated for more general uses within commercial 3D printing and represented a novel transfer of applications for certain of these methods.

What was the impact

The intelligence provided by the BCGC team was incorporated into an internal initiative at Autodesk subsequently named Project Nido, which sought to integrate sustainability, green chemistry and biomimicry into Autodesk's product design and development activities. Autodesk published a series of digital articles about AM, green chemistry and bio-inspired design on their community of practitioners web platform and began discussions with the BCGC team about incorporating chemical hazard assessment (CHA) into the databases being developed for building information management (BIM) software. Additionally Tom McKeag published a chapter in volume 10 of the Yale University Press' Green Chemistry Handbook series, Tools for Green Chemistry entitled "Shaping the Future of Additive Manufacturing: 12 Themes from Bio-Inspired Design and Green Chemistry" in 2017 (McKeag, 2010). The chapter was based on the bio-inspired design and green chemistry concepts for AM outlined in the unpublished 2015, consulting report.

This project also set the stage for subsequent work in the subject. The initial consulting work instigated the formation of a core group of researchers who would continue the investigation over the next several years. Autodesk engaged the fall 2015, Greener Solutions graduate course to continue needed research, and funded a subsequent industry roundtable with the nonprofit Northwest Green Chemistry, and cofunded with BCGC an extended internship for a key member of the consulting team, chemist Justin Bours.

Phase two: the fall, 2015 Greener Solutions course

The initial consulting report and subsequent publication drafts were well received at Autodesk and the company agreed to partner with the Fall 2015, Greener Solutions graduate course at UC Berkeley to investigate more sustainable SLA printing and material options based on nature.

What was the challenge

The initial consulting report had characterized the relative chemical hazards of typical resins, including a baseline for the target formula, PR48, provided a dozen broad strategies from nature for consideration when searching for alternatives, and had spotlighted three examples from biomedical tissue engineering that had demonstrated some limited success with some of these concepts. The Greener Solutions team was charged with building upon this foundational intelligence, pursuing some of the nature-based strategies across the full spectrum of the SLA printing process (looking at it as a system), and proposing a spectrum of actionable recommendations for adjusting or replacing the current PR48 formula (Table 5.4).

Why this project was important

With the initial hazard assessment of PR48 and commonly used resin components now complete, Autodesk was looking for specific interventions it could pursue to redesign or replace the PR48 formula. In keeping with its continued commitment to sustainability leadership, the company hoped that results from some innovative solutions offered by the Greener Solutions team could be shared with the larger AM community. This was in alignment with their policy approach of open-sourcing the details of the Ember printer and the formula for PR48. Typically, formulators in the industry do not reveal their formulas, since these formulas are key profit-making intellectual property. An assessment of a known formula and discussion of possible alternatives was an important first step toward greater transparency in the industry.

Who were involved

The Greener Solutions students who prepared the final "opportunities map" report and presented to Autodesk were: Chen Cheng (chemistry), Ann Dennis (architecture), Lee Ann Hill (public health), Coleman Rainey (physics), Brian Rodriguez (public health). The course was taught by BCGC staff and students: physician and School of Public Health faculty Megan Schwarzman, M.D.; chemist and BCGC executive director Dr. Martin Mulvihill; bio-inspired design expert, and later executive director at BCGC, Tom McKeag; chemist Dr. Justin Bours; and chemist and postdoctoral student Dr. Heather Buckley. Autodesk partners included Susan Gladwin, Dawn Danby, Brian Adzima, Chris Venter, and Shalom Ormsby.

What was the solution

After 16 weeks of research, the students in the 2015, Greener Solutions class presented Autodesk with three different approaches to the challenge of redesigning the PR48 resin. The students further identified five potential solutions, targeting three different aspects of the PR48 formulation (Fig. 5.4).

The first solution proposed a drop-in functional substitution: replacing the photoinitiator compound with a natural substance such as curcumin or riboflavin, which have significantly better hazard profiles than the photoinitiator currently in the PR48 formulation.

The second was a recommendation to replace the backbones being modified by the acrylates in the resin, attaching them to triglycerides or chitosan, thereby

TABLE 5.4 Components of PR48.

	Ingredient	%	Function	Hazard(s)
Reactive Oligomers	Allnex Ebecryl 8210	39.78	Crosslinking	Reproductive & Developmental Toxicant Skin Sensitizer Skin & Eye Irritant Aquatic Toxicant
	Sartomer SR 494	39.78		
Reactive Monomer	Rahn Genomer 1122	19.88	Crosslinking	Skin & Eye Irritant Aquatic Toxicant
Photoinitiator	Esstech TPO+	0.4	Polymerization: Initiation	Reproductive Toxicant Skin Sensitizer Aquatic Toxicant
UV-blocker	Mayzo OB+	0.16	UV light penetration control	Skin & Eye Irritation Persistent

From: Cheng et al., 2015. Autodesk and stereolithography 3D printing. Bio-inspired resins for better human and environmental health unpublished report (Table 2.1).

FIGURE 5.4 Strategic approaches to SLA resin alternatives. *From: Cheng et al., 2015.* Autodesk and stereolithography 3D printing. Bio-inspired resins for better human and environmental health, *unpublished report.*

reducing the hazard posed by the acrylate monomers and to increase the molecular weight of any cross-linking acrylates. Understanding that molecular weight is tied directly to viscosity and viscosity directly affects print speed, the team recommended adjusting the fatty acid chains or adding unspecified diluents to the eventual formulation in order to match liquid flow with the operation of the printer. This second solution took inspiration from the structure of fatty acids and the chemistry of chitin formation in insect exoskeletons.

The third and most disruptive of the solutions that the students presented took inspiration from the mechanisms by which oyster shells and human bone tissues are constructed, and outlined a process wherein a UV-activated compound in the resins would alter the pH of the resin, inducing polymerization or the formation of Metal-catechol complexes. The report suggested a combination of either phenylgloxylic acid or ketoprofen as anions and either 1,5-diazabicyclo[4.3.0]non-5-ene or phenethylamine as cations to form the quaternary ammonium salts that would then act as the photoinitiated base generators. The students identified these compounds as having both desirable performance properties and more favorable human and environmental health properties than the components of the extant PR48 resin. They acknowledged, however, that pH-based SLA processing technology required significant additional research and development before it was likely to become viable for the AM market (Fig. 5.5).

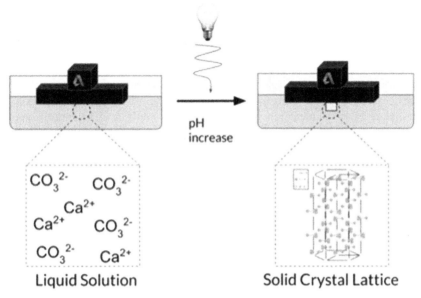

FIGURE 5.5 Concept diagram of calcium carbonate in liquid solution transformed to a solid crystal lattice by light and pH increase. *From: Cheng et al., 2015*. Autodesk and stereolithography 3D printing. Bio-inspired resins for better human and environmental health, *2015 unpublished report.*

What was innovative about the solution

First, the public assessment of the relative hazards of a known SLA resin formula was novel and ground-breaking in and of itself and proposing substitutes was the pioneering start to a more transparent and public discussion about chemical safety within the industry. Autodesk can be proud of their role in facilitating and promoting this long-range view of commercial success and community caring.

The materials solutions outlined by the Greener Solutions team were innovative in several ways. The repurposing of natural materials as photo-initiators, the proposed modifications to the acrylate subcomponent of the SLA resin, and the identification of a novel method of pH-based polymerization/complexation, provided a range of potential interventions that Autodesk could implement depending on available resources and expertise.

Second, the range of solutions offered by the Greener Solutions team included the use of bio-based feedstocks and photoinitiators that were of easier adoption with well-known materials and processes. The use of bio-based materials such as curcumin or riboflavin as drop-in substitutes for the PR48 photoinitiator was the simplest, although some additional work would be needed to optimize formulation and production using these alternatives. The bio-inspired use of triglycerides or chitosan as backbones for acrylate-based resins could produce an economical and effective chemical solution that did not deviate significantly from a current formulation, reducing the degree of development necessary for incorporation into existing product lines.

Third, the team's solution set also included the disruptive idea of changing the activator within the SLA print system to a PH-based photoinitiator, and they described two mechanisms to achieve this. The pH-regulated photo-resins described in the student report, once perfected, could lead to entirely new methods of AM, lead to additional technologies, and potentially novel fields of research. Applications of pH-regulated photo-resins could extend well beyond the field of AM into construction, scientific research, and even medicine.

What was the impact

The results of this investigation were well received by the collaborators at Autodesk but not acted upon by upper management at the company. Indeed, Autodesk was going through a significant period of transition that included the stepping down of its president in February 2017, the shift to cloud-based information subscriptions and downsizing to core business activities that would soon eliminate the sustainability sector, the Spark (Ember) program, and most of the positions that had collaborated with BCGC.

The research and creative ideas developed in this phase would, however, form the partial basis for much of the subsequent work outlined below in phase five and six. In particular, the third, disruptive innovation option of employing a pH change to initiate polymerization would lead to further development of more sustainable options.

Phase three: developing a framework for more sustainable material choices in additive manufacturing at Autodesk and Northwest Green Chemistry

What was the challenge

In late 2015, the sustainability section at Autodesk requested of BCGC that a methodology be developed that could define, in a rational and quantifiable way, what it meant to be a "biofriendly" AM resin; in other words to have little or no negative impact on human and environmental health. BCGC and Northwest Green Chemistry (NGC), a non-for-profit sustainable chemistry consulting firm in Seattle, WA, collaborated to share financial and technical support for BCGC associate and chemist Justin Bours to intern at NGC and Autodesk and continue the development of this framework.

From late 2015 to early 2016, Bours worked with Julia Cabral, the Global Environmental Health and Safety Manager, and others on the original phase one team at Autodesk, to develop a holistic AM sustainability framework that combined elements of chemical hazard assessment (CHA), alternatives assessment (AA), and life cycle assessment (LCA). The purpose was to support stakeholders in making wiser and easier materials and process choices. This framework was published first as a white paper in early 2016, and then in the Journal of Industrial Ecology in 2017 (Bours et al. 2017).

Why this project was important

This work was important for several reasons. First, it was aligned with the wider missions at Autodesk in sustainable design and gaining greater participation in the AM market. It also aligned with NGC's mission to educate and provide tools to the public and industry about wiser material and process choices that affect public health. Second, it provided new metrics for Autodesk and other stakeholders in the industry sector to make judgments about material selection. Third, it formed the basis for the wider education and engagement of professionals first at Autodesk, then at several industry presentations and professional conferences, and finally for the stakeholder roundtable outlined below in phase four.

Who were involved

Dr. Justin Bours, sponsored by BCGC and interning at both Autodesk and Northwest Green Chemistry (NGC), Julia Cabral, Global Environmental Health and Safety Manager, Dawn Danby, Senior Sustainable Design Program Manager, and Susan Gladwin, Sustainability Project Manager at Autodesk, and Lauren Heine, executive director at NGC. Tom McKeag, now executive director at BCGC, provided overview and administrative support.

What was the solution

The team developed a framework that complemented LCA with hazard and green design metrics derived from analyzing human health and environmental impacts in the later stages of the AM life cycle. They identified suitable existing methodologies for evaluation across these stages and synthesized the methodologies into higher-level metrics for comparative analysis of materials. With these metrics in place, they compared two common AM materials, Autodesk Standard Clear Prototyping Resin (PR48), an open-source formulation used in photopolymerization processing AM, and *bio*−polylactic acid, a ubiquitous, biosourced polymer used in an extrusion-based AM system called typically fused filament fabrication. One of the key design concepts in this project was to use a bundle of existing assessment tools, but customize each selectively at appropriate material life stages in order to satisfy a wide range of sustainability goals (Fig. 5.6 and Table 5.5).

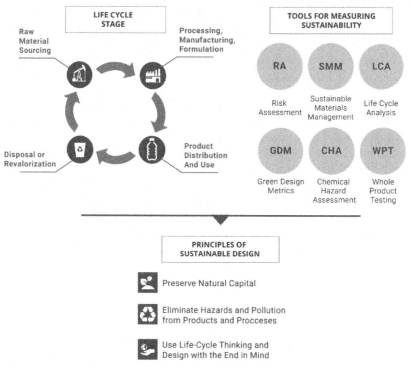

FIGURE 5.6 Sustainable design strategy using a variety of measurement tools at appropriate life cycle stages. *Adapted from: Justin Bours (2015), unpublished image.*

What was innovative about the solution

The AM materials selection framework developed by Justin Bours et al. was novel in that, first, it combined tools from LCA, AA, and CHA, allowing users to generate more comprehensive materials selections. It came from an approach that held that no one tool could adequately measure the breadth of sustainability, and that each life cycle stage should inform which sustainability tools to use. Most LCA assessments at the time did not include a robust scrutiny of inherent chemical hazard, but instead focused more on material and energy costs. This framework addressed that shortcoming in a new way. Second, other types of assessment tools aligned with the appropriate life cycle stage were synthesized into a higher level score in a novel approach yielding a more rounded characterization of a material (Fig. 5.7).

What was the impact

The new method of combining toxicological data with a range of other sustainability protocols within a customized life cycle intervention scheme had now been "test driven" in a complete scoring cycle for two types of AM

TABLE 5.5 Material toxicity assessment at print stage for Pr48 and PLA.

From: Justin Bours (2015), unpublished image.

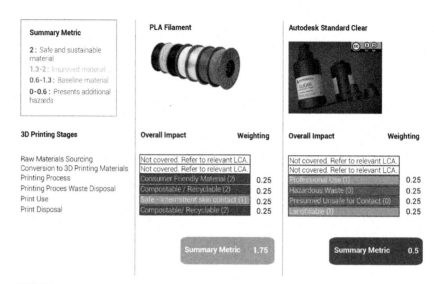

FIGURE 5.7 Test case summary of biofriendly framework. *Adapted from: Justin Bours (2015), unpublished image.*

materials, PR48 used in SLA, and PLA used in fused filament processes. A logical relative ranking for two materials was now demonstrated and would form the basis for later refinement.

During this phase Bours et al. published *Addressing Hazardous Implications of Additive Manufacturing: Complementing Life Cycle Assessment with a Framework for Evaluating Direct Human Health and Environmental Impacts,* in the Journal of Industrial Ecology, May, 2017 (Bours et al. 2017). Bours and others were also able to present the project at several national and regional gatherings and start some of the earliest industry discussion about the need for a full-cycle materials review at the chemical hazard level.

Most importantly, although hard to quantify, this phase would allow one researcher to think deeply and to test out a scheme for harmonizing assessment tools and data, expose data gaps, and attempt to reconcile conflicting metrics, relative usefulness, and clarity of communication that would be needed for the framework to be employed by the industry.

This initial framework would form the basis for the next phases of this ongoing investigation: an industry round table to garner feedback from stakeholders (phase 4), and, ultimately, a scorecard that could be used to rank materials (phase 5).

Phase four: industry roundtable

What was the challenge

Autodesk recognized the need to garner wider industry feedback for any new assessment tool, so asked Northwest Green Chemistry (NGC) to create and conduct a series of roundtables. The purpose was to gather additional input from stakeholders in the AM industry and refine further the sustainability framework. This proposed collaboration, led by NGC founder Lauren Heine, was successfully funded, leading to the creation of "The Sustainable Materials Roundtable in Additive Manufacturing" which comprised three meetings over the course of 2017, and early 2018, and included representatives from Cradle to Cradle, and Dartmouth College in the organizing effort. The objective was to inform the development of a "practical and useful framework to help identify safer and more sustainable AM materials and processes."

Why this project was important

This was the logical next step in turning this prototype framework into a workable tool. The existing model for the assessment framework by Bours was novel in its approach and much more comprehensive in its materials assessment than many existing tools (due to its addressing inherent chemical hazard). Its usefulness to working professionals, however, had not been tested, and team members recognized that the framework had a lot of interconnected parts and subtleties that might be daunting to those not familiar with toxicology, and chemistry. Moreover, the framework had been developed within only one organization, Autodesk, so may have missed the mark on the decision-making needs and concerns of other industry stakeholders involved in 3D printing or materials assessment.

Who were involved

Justin Bours, now of Cradle to Cradle, Lauren Heine, Northwest Green Chemistry, Amelia Nestler, Northwest Green Chemistry, Mark Buczek, Northwest Green Chemistry, Jeremy Faludi, Dartmouth College, Tom McKeag, the Berkeley Center for Green Chemistry (BCGC), and approximately 40 international representatives from material and chemical manufacturers, AM users, printer and software makers, consulting firms, NGOs, academia, and government.

What was the solution

The round table organizers were keen to get feedback on the main question: "What type of tool would be most useful to you as a decision-maker in assessing materials and making a choice about whether and how to use them?" This overarching question would lead to three productive meetings and follow-up polling that teased out several important considerations for developing the tool.

Initially polling revealed that all were agreed that a proper tool would: address both material selection and product design; that no one of the currently available assessment protocols (life cycle analysis, chemical hazard assessment, etc.) would be sufficient to fully measure sustainability; that any tool should be simple, visual, and easy to use; that tradeoffs were inevitable, but they should be clearly identified within the tool for good transparency of methodology.

Nevertheless, basic questions remained about scale, points of application and what criteria to include in the scoring. For example, would the tool be used to assess a whole product or just single materials? Could you realistically apply it to material across process technologies? Should you include more variables beyond sustainability that were typically driving decision-making, like cost?

The organizing team responded by proposing four types of prototype models and proffering the pros and cons of each to the invitees in the next phases of the roundtable series. The types were:

Alternative assessment (AA)

AA is a process for identifying and comparing chemical and nonchemical substitutes to replace chemicals or technologies of high concern. Several well-established models for AA had been produced by the Interstate Chemicals Clearinghouse (IC2), the National Research Council, the State of Washington Department of Ecology, the State of California Department of Toxic Substances Control, and the company Method. The team noted that while this method typically offered a formal and comprehensive framework for decision-making and included often under-represented variables, it lacked support on making choices, clear, simple graphics, or transparency in how tradeoffs were made (Fig. 5.8).

Augmented life cycle analysis (LCA)

Typical LCA software like ReCIPe lacked robust assessment of health impacts, so a model was offered that would augment an LCA-based program with enhanced chemical hazard and toxicological assessment. This created a new customized endpoint for the program that would identify chemicals of concern through the full cycle, screen for toxicological issues like CMRDEs, sensitizers, neurotoxicants, and PBT; break out by exposure pathway, and then rank for hazard, exposure and risk. While this approach presented one normalized metric that combined LCA and health and was simple and fairly transparent, it was unclear if it would sufficiently discriminate within the many different AM printing situations and offer decision-making support (Fig. 5.9).

Product Design Scorecard

This model was based on one developed by Jeremy Faludi that was originally focused on enumerating beneficial design traits and ranking them as they might relate to sustainability impacts. The model would be applied to both print technologies in printers and the materials used. While simple to use and

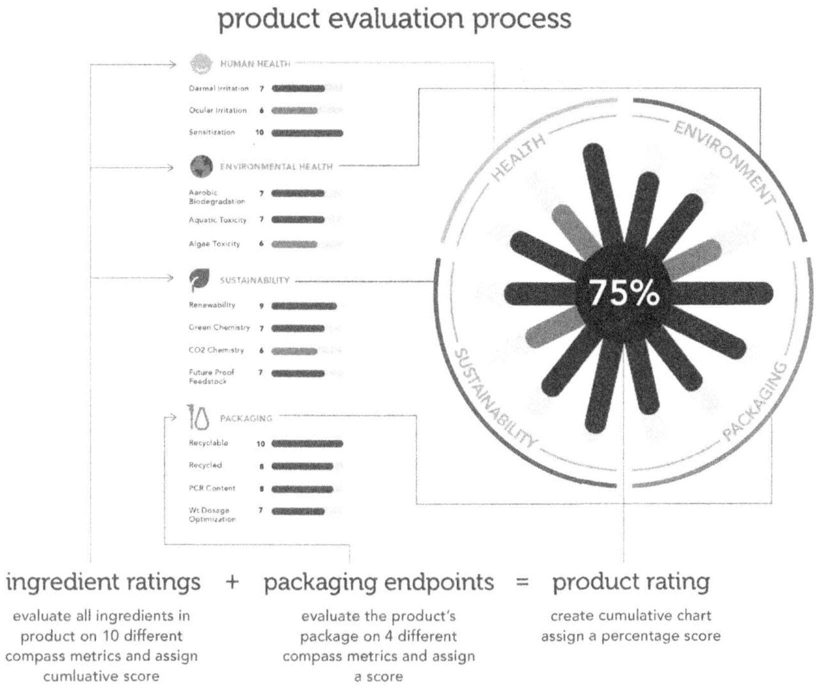

product evaluation process

ingredient ratings   +   packaging endpoints   =   product rating

evaluate all ingredients in
product on 10 different
compass metrics and assign
cumluative score

evaluate the product's
package on 4 different
compass metrics and assign
a score

create cumulative chart
assign a percentage score

FIGURE 5.8 Prototype one, alternatives assessment example: Method's compass of clean. *Adapted from: https://methodhome.com/benefit-blueprint-compass/.*

FIGURE 5.9 Prototype two: augmented LCA example. *Adapted from: Northwest Green Chemistry (2018), unpublished image.*

understand, this model did not provide a quantitative measure of impacts, and ranking criteria and tradeoffs were unclear. Each module would be visualized separately and the user would view each as a separate score to use in his/her own weighting, perhaps in bar graphs or so-called spider diagrams. This approach was comprehensive, transparent about tradeoffs and gave some aid to decision-making. It required quite a bit of data for the analytics, however, and did not appear to address some of the specific needs of product designers (Fig. 5.10).

Hybrid model

This model was most like the Bours et al. framework published in the Journal of Industrial Ecology. It combined LCA, health and reutilization metrics to yield a combined score for sustainability. Each of the three modules would have its own criteria, metrics and templates for each of the life cycle stages. For example, within the health module, ultrafine particle emissions (PM 2.5) might be chosen as one of the important criteria, the tools for assessment chosen might be CHA and Risk Assessment (RA), and the system of metrics might come from the Cradle to Cradle material health and exposure assessment methodology (Fig. 5.11).

The results of the roundtables were that the hybrid model and the scorecard approaches were preferred and should be combined with the best characteristics of each retained. It also became clear that a balance must be achieved within the form of the tool for the following contradictions: comprehensive data collection versus user overload; control versus freedom on user decision-making; ease of use and scientific rigor; range of application and precision of intelligence.

A design-guide checklist/scorecard may list beneficial design traits and score them based on their ability to improve impacts.

		HIGH priority (3 pts. each)	MEDIUM priority (2 pts. each)	LOW priority (1 pt. each)
Printer Design		Design to minimize idle time (ease of sharing, minimal setup and clean-up time)	Low-energy printing process (chemical bonding not melting)	Design to minimize material in prints
		Design to encourage bed packing for photopolymer, inkjet, and laser sintering printers	Design to minimize support material (without print failure)	Design to minimize material waste
		Automatic low power standby (<1% of operating power)	Energy-efficient equipment systems, (insulation, motors, electronics)	Design to minimize failed prints
Printing Materials		Non-toxic, compostable photopolymers for SLA, DLP, polyjet, and CLIP printers		Infinitely reusable metal powder produced from recycled material
		Chemical bonding (not melting) of compostable biomaterials		

FIGURE 5.10 Prototype three: product design scorecard example. *Adapted from: Northwest Green Chemistry (2018), unpublished image.*

Prototype 4: Hybrid Method

Spider diagrams could be used to visualize scores for each stage of the life-cycle (a, b, c, d, e, and f) for all three modules

- Could also be represented by bar charts
- Decision-making schemes not inherent to diagrams, but could be implemented

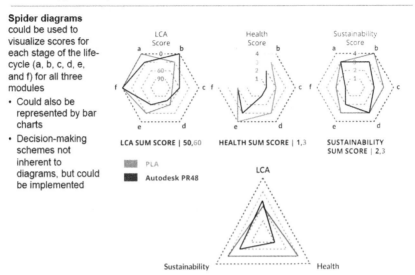

FIGURE 5.11 Prototype four: hybrid example. *Adapted from: Northwest Green Chemistry, (2018), unpublished image.*

Overall, respondents desired a tool that could be used by customers to guide purchasing decisions and by producers to guide materials and product design. They also saw scoping as a challenge and that combining all these tools creates a potentially unwieldy number of endpoints to consider. Moreover, it remained difficult to compare endpoints across significantly different printing technologies. Finally, it was stressed that the tool must be designed to be accessible to users with a range of backgrounds and levels of expertise.

What was innovative about the solution

The AM industry did not have any specifications for chemical safety for materials used in many of their technologies, so in a broad sense, this endeavor was ground-breaking by formally opening the discussion and convening a collaboration to inform the design of a useable tool for the sector.

This was the first time, to the team's knowledge, that a purposely balanced group of stakeholders from the AM sector had been brought together to review and provide input to the design of a chemical hazard/sustainability assessment prototype tool. This was also the first time that several existing assessment tools had been combined and customized to material life cycle stages in AM.

What was the impact

From a content perspective, this roundtable provided invaluable insights into the foundational contradictions to be encountered in designing a technical assessment tool for a wide range of users. The series also accelerated the development of the framework as a result of the collaborative discussions and brainstorming. Most immediately, it helped the informal development team chart a course for a tangible product. As will be seen in the next phase, five, the roundtable was also the germination of a more formal consolidation of the development team and study leading to that product.

Finally, it should be noted that information flow at the roundtable went two ways, in that many of the invitees were informed about material choice issues that they perhaps were not aware of, and, in the course of the discussions, some new ideas were entertained that might not have surfaced without the meetings. This served to further the goal of educating and influencing decision-makers toward identifying and choosing safer materials and processes.

Phase five: designing a scorecard for chemical hazard assessment: the P2 project

The Berkeley Center for Green Chemistry continued the work toward safer SLA resins in September 2018, after being awarded a grant from the US EPA's Region 9 Pollution Prevention (P2) program. The P2 grant program fosters industry collaboration to avoid pollution discharges by the adoption of safer practices in manufacturing. BCGC had proposed collaboration with chemical manufacturer Millipore Sigma, nonprofit Northwest Green Chemistry (NGC), chemist Justin Bours of Cradle2Cradle, and mechanical engineer and sustainability expert Jeremy Faludi of Dartmouth College and later Technical University of Delft. The 3-year project would survey the current state of SLA formulations and ingredients, develop a hazard scorecard system (building on the work in the phase four round table), search for alternatives, compare an existing baseline formula PR48 with proposed alternatives using the chosen scorecard, and test selected alternatives for technical performance. This work has been completed and we will discuss the results of the development of a new comparative CHA scorecard and user guide. BCGC also advised a graduate student team in the fall, 2019, Greener Solutions course in a search for alternatives to acrylate cross-linkers, and the results of that work is described in phase six.

What was the challenge

This overall project was titled "Developing and Performance Testing a Safer Formulation for Stereolithography Printing Resin." The stated purpose of this project was to produce and disseminate a replicable best practices methodology for chemical assessment of SLA printing resins, a set of standards for formulation of safer resins, and actual alternative formulas that could be tested in the field with AM players brought together in an industry roundtable.

The results of the work toward the first part (compiling of best practices for chemical assessment of the SLA printing resin) will be discussed here. The challenge for this part was to produce a defensible set of criteria for judging individual chemicals and formulas for both human and environmental health, and to create a useable scorecard that organized these criteria into a tool for decision-makers to rank materials.

Although the P2 team would build on the progress made in the previous framework development and roundtable phases, this project had a focus on chemical hazard of materials and substitutions that would avoid pollution. Therefore, some of the sustainability criteria developed earlier were not included in the quantitative scorecard produced. Nevertheless, the team continued to address all of the previous issues of scorecard design: data gaps, unreconciled metrics, the contradiction between comprehensive data and user friendliness. The team's communication of results strategy comprised three widening circles of content, starting with the core focus on a quantifiable comparison of chemical hazard of alternatives using a defensible and replicable methodology; framing those results within a context of sustainability factors; and finally, and to a much lesser extent, making mention of wider societal concerns such as social justice and public health burdens.

Why this project was important

Several years on from the start of BCGC's engagement with the safety of chemicals used in AM, the industry had not established any EHS performance standards for chemicals in resins. At most, some commercial operations had developed best practices for handling and disposal of material and the major standards organizations were still forming committees to set preliminary standards. The most advanced of these activities were associated with air emissions from technologies like fused filament deposition.

The development of criteria incorporated into a useable scorecard, therefore, would be a major step toward industry engagement and hopefully adoption. Designing formulas based on safer EHS standards in the first place would be a radical change and model for the fast growing sector. If the formulas passed the technical performance bar, then they would offer a tangible alternative and prompt greater change.

Who were involved

BCGC Executive Director Tom McKeag, Senior Researcher Dr. Tala Daya, and Researchers Dr. Rachel Scholes and Nicole Panditi joined Drs. Lauren

Heine and Anna Montgomery of NGC, Dr. Jeremy Faludi of Dartmouth College and later Technical University of Delft, and Dr. Justin Bours, consulting chemist, in this project. Early collaboration and technical guidance on bio-inks were provided by Drs. Samy Punnosamy, Ganga Panambur, and Didarul Bhuiyan of Millipore Sigma's St. Louis, Missouri and Milwaukee, Wisconsin branches. Later phases included important collaboration and information sharing with the nonprofit ChemForward, 3D printing consultant Dr. Mike Idacavage, the industry association Radtech, and the National Institute of Standards and Technology (NIST).

What was the solution

The overall strategy of part one of this investigation was to develop an assessment scorecard for inherent chemical hazard assessment (CHA) that could be used by companies to make safer material choices. This scorecard would give a single numerical score for each chemical, based on a bundling of hazard assessments for human and environmental health. Job one was to frame the boundaries of performance criteria (scope of work) and identify the target AM technologies and potential users (Table 5.6).

It became apparent immediately that the process and material differences of the various AM technologies would have to be sorted out in order to establish a reasonable scope of work. This was done with a simple spreadsheet comparing components of printing systems: activators, monomers and oligomers, and additives like colorants and UV blockers. The researchers noted the shared characteristics of the different systems and narrowed the scope to include three technologies: SLA, extrusion, and inkjet processes. All shared a commonality in the activation process; the use of ultraviolet light, either in the polymerization of formulas or postcuring. Typical materials were recorded and any data on technical and EHS performance. A complementary investigation of bio-inks was also performed, aided by information from research scientists at industry partner Millipore Sigma. The P2 researchers reasoned that a wide, divergent search for alternative materials across the three technologies and in bio-ink development could yield safer material candidates that might be applied to SLA.

Translations of materials across these technologies would not be necessarily straightforward. Examples of translation challenges were the difference in performance viscosity and print speed. Material/process combinations that were candidates were reviewed for technical performance by using industry standards for criteria like print speed and mechanical strength, and underperforming candidates set aside as lower priorities rather than discarded, with the reasoning that some failings could be mitigated with augmented processes or additional materials. A simple generic solution example is the addition of microfibers into plastic mixes to improve tensile strength.

TABLE 5.6 Additive manufacturing life cycle stage health impact comparison.

	inputs	processes	exposed parties	routes of exposure	hazards or impacts
1 Raw Material Sourcing	· energy · fuel for transport · process chemicals	extraction of raw materials	worker environment	inhalation, dermal environmental — ground water, air	Chemical hazards Chemical hazards GHG emissions
2 Chemical Manufacturing	· energy · fuel for transport · water use · process chemicals · precursor chemicals · formulation chemicals	chemical manufacturing	worker	inhalation, dermal	chemical hazards
3 Material Formulation	· energy · fuel for transport · water use · process chemicals · formulation chemicals	formulation	environment	environmental ground water, air, energy use	Chemical hazards GHG emissions
4 Print Processing	· energy · formulation chemicals · process chemicals		environment printing operator	environmental (energy use) inhalation, dermal	GHG emissions Chemical hazards UFP emissions VOC emissions
5 Disposal and Reutilization	· energy · formulation chemicals · print material		environment	environmental (ground water)	Chemical hazards GHG emissions waste
6 Print Use	· formulation chemicals · print material		print user	dermal, oral	chemical hazards
7 Disposal and Reutilization			environment	environmental (ground water)	Chemical hazards waste

From: Justin Bours (2018), unpublished work for the Berkeley Center for Green Chemistry.

While the core focus of the project was on the chemical hazard of ingredients, it also became apparent quickly that the team would have to distinguish between process chemicals and formula chemicals and map where in the life cycle of a 3D printed object these chemicals were used or produced as waste or biproducts. Identifying this life cycle context could now be used to judge relative impact, narrow the core assessment scope, and discuss wider implications of some of the chemical hazard assessment (CHA).

With relative priorities for the project's data collection set, and the scope narrowed to type of technology and life cycle phases, Justin Bours began a chemical inventory, determining what chemical classes were relevant, classifying them for their inherent chemical hazards, and establishing the metrics for translating hazard data into numerical ratings.

The hazard data for the relevant chemical classes were gleaned from Material Data Safety Sheets, scientific and toxicological literature, regulatory lists and databases, other chemical hazard databases, or directly from the manufacturers. Next the hazard data were sorted by the 16 hazard endpoints derived from GHS (Globally Harmonized System)-based protocols, like Green Screen, and Cradle to Cradle's Material Health Assessment Methodology.

Dr. Bours then developed a chemical hazard scoring system for both human and environmental health endpoints that rolled up the 16 standard endpoints of GHS-based assessment methods into four hazard groups for Human Health (CMRDE, Oral/Dermal/Inhalation Toxicity, Sensitization, Irritation), and four hazard endpoints into three hazard groups for Environmental Health (Acute Aquatic Toxicity, Chronic Aquatic Toxicity, Fate) (Table 5.7).

An analysis of data gaps and confidence levels followed. A data gap was discounted if reasonable comparable data from another endpoint existed; otherwise, when rolling up hazard endpoint ratings into hazard group ratings, any remaining data gaps were translated as a range of worst possible hazard group rating to best possible hazard group rating.

Bours assigned Hazard Group Ratings for each chemical, presenting them as a range between the worst possible hazard rating (1) for that hazard group and the best possible hazard rating (4) for that group. The best hazard rating for a hazard group was based on the worst hazard rating for a toxicity endpoint within that hazard group. If there was no available data for that hazard group then the best hazard rating was automatically rated as a "4." The worst hazard rating for a hazard group was based on the worst hazard rating for a toxicity endpoint within that hazard grouping, or automatically rating it a "1" if there was a data gap for a toxicity endpoint that could not be discounted. Team review discussions concluded that this range of ratings was the most direct approach to indicate confidence level in the data results: wider ranges indicating more uncertainty, while still showing relative hazard among chemicals tested.

Each of the hazard groups were given weighting coefficients, based on relative danger: cancer, for example, being more directly concerning than skin

TABLE 5.7 Chemical hazard scoring for human and environmental health.

From: The Berkeley Center for Green Chemistry (2021), unpublished report.

irritation, and the combined hazard group scores, now weighted, combined for a single score for human and environmental health hazard.

With the scoring for individual chemicals in hand, the team could now turn toward how to score formulations made up of these individual chemicals. First, Bours established rules for immediate elimination of formulas from the scope of work due to extreme hazard potential, and for red-tagging potentially very hazardous formulations. Next the team discussed and agreed upon a weighting system for formulas, using the individual chemical scores and weighing by the relative percentage of physical weight in the formula. Once again, this team discussion considered the balance between precision and straightforwardness for the user. Human and Environmental health hazard scores were, finally, combined to arrive at a single numerical score for a formula (Table 5.8).

What was innovative about the solution

This scorecard system was devised to correct perceived shortcomings in current CHA strategies that use benchmark categories to rate chemicals. The intent was to provide more nuanced profiling of individual hazard groups so that a more precise picture of hazard could be available to those who needed it, but also provide a single, bundled score to decision-makers who were looking for more general guidance. Moreover, this scoring system, in its weighting methods, allows for exposure impact and therefore takes into account life cycle phases as well as inherent chemical hazard. Finally, this system shows, at a glance, both relative scores, and, because they are shown in ranges, relative uncertainty, and data confidence level.

For the first time ever, the BCGC team developed a comparative CHA schema for AM resins that aggregated and streamlined standard GHS (Globally Harmonized System) toxicity endpoints into one measurement or rank. This schema aggregated traditional GHS hazard endpoints, reconciled different rating systems from various existent scoring systems, assigned numerical ratings for new hazard groups, and rolled up ratings into a single chemical hazard classification for both human and environmental health, before assigning an overall, weighted score range based on best and worst case scenarios.

Using this schema, the team designed and developed a comparative CHA scorecard for components in AM resins. A user can perform a five-step comparative hazard analysis, rank individual chemicals of concern and formulations and compute human health, environmental, and overall scores for their target chemicals. Overall scores are represented as a range of values, based on the individual best and worst scores for constituent factors. Formula scores are computed by the percentages of individual components by weight. A user can also customize this scorecard.

To support this tool the researchers developed a step-by-step user guide with examples and an appendix that guides the user to important sources of data with

TABLE 5.8 Overall chemical hazard scoring template with example.

AM Formulations	Direct Human Hazard		Environmental Hazard		Overall Chemical Hazard Score	
	Worst Case	Best Case	Worst Case	Best Case	Worst Case	Best Case
	0.5		0.5			
PR48	3.20	3.20	3.83	3.84	3.52	3.52
Acrylate CastorOil	0.00	0.00	0.00	0.00	0.00	0.00
Formulation 1	0.00	0.00	0.00	0.00	0.00	0.00
Formulation 2	0.00	0.00	0.00	0.00	0.00	0.00
Formulation 3	0.00	0.00	0.00	0.00	0.00	0.00
Formulation 4	0.00	0.00	0.00	0.00	0.00	0.00
Formulation 5	0.00	0.00	0.00	0.00	0.00	0.00
Formulation 6	0.00	0.00	0.00	0.00	0.00	0.00
Formulation 7	0.00	0.00	0.00	0.00	0.00	0.00
Formulation 8	0.00	0.00	0.00	0.00	0.00	0.00
Formulation 9	0.00	0.00	0.00	0.00	0.00	0.00
Formulation 10	0.00	0.00	0.00	0.00	0.00	0.00

Comparison of Scores

PR48, 3.52
PR48, 3.52

Overall Hazard Score (1-4)
4.50
4.00
3.50
3.00
2.50
2.00
1.50
1.00
0.50

● Worst   ● Best

The chart above plots the best and worst scores for each formulation (#1=Formulation 1). Higher scores (max = 4) represent greater safety.

From: The Berkeley Center for Green Chemistry (2021), unpublished report.

which to populate the spreadsheet-based, toxicity endpoints, read-across scorecard. This appendix also includes easy-to-read binary decision trees to guide the user in making scoring choices from curated lists. We made the scorecard and user guide available to a select group of beta testers on this Google Drive site and gathered qualitative insights from a small cohort of testers.

What was the impact

This new scorecard system has established a novel set of metrics to review the relative chemical safety of SLA formulas, and could serve as a model for the review of other materials. The scorecard was shared with the nonprofit ChemForward and served as a background document in the development of their web platform for a universal comparative toxicology testing system. This platform contains portfolios of AAs available to industry.

The draft scorecard was presented by the team to make the case for green chemistry at the resin formulation stage and safer alternatives to national audiences of researchers, industry trade associations and government officials, including Radtech, and the National Institute of Standards and Testing (NIST). Nearly all previous health and safety focus within the national industry trade network had been on regulation compliance and safe handling of hazardous materials practices.

Phase six: the fall, 2019, Greener Solutions course

What was the challenge

In the fall of 2019, a second team of Greener Solutions students took up the SLA challenge, working to find additional alternatives to existing SLA resins. In partnership with researchers in the Berkeley Center for Green Chemistry, Mechanical Engineering at UC Berkeley, and Dr. Justin Bours at Cradle to Cradle, the team aimed to identify innovative directions for development of SLA resins that would reduce or eliminate exposure to the cross-linking agents (meth)acrylates.

Why this project was important

Finding safer alternatives to these commonly used cross-linkers would avoid a range of environmental and human health problems for workers and consumers. Acrylates and methacrylates comprise approximately 99% of the PR48 formula used as a known baseline and exhibit toxicity due to reactions with biomolecules. In particular, the alpha,beta-unsaturated carbonyl present in (meth)acrylates is electrophilic, such that it reacts with and depletes nucleophilic biomolecules including thiol-containing proteins (e.g., glutathione). The toxicity of electrophilic compounds depends on their reactivity with these proteins: reactivity with glutathione and other thiol-containing biomolecules has been linked to health endpoints observed for (meth)acrylates, such as skin and respiratory sensitization and aquatic toxicity (Chan & O'Brien, 2008). Therefore, although the α,β unsaturated carbonyl is a useful chemical feature for cross-linking reactions, new cross-linking mechanisms will be necessary in order to avoid the inherent toxicity of electrophiles via reactions with biomolecules.

The Greener Solutions 2019, final report characterized the human health issues associated with acrylates and methacrylates:

The main hazards associated with 3D-printing resins are caused by the presence of the acrylate groups. If absorbed into cells, acrylates can intercalate into the DNA through conjugate addition, rendering acrylates potential mutagens, as well as reproductive and developmental toxins. In light of their hazard profile, many acrylates are also regulated under California Proposition 65. *In addition, acrylate monomers in the current PR 48 resin also exhibit moderate to severe aquatic toxicity, and requires that users properly dispose of the resins and rinse waste. The main route of exposure to acrylate is through skin absorption during monomer production and 3D-printing. Because the monomers have high boiling points and low vapor pressure at room temperature, inhalation of volatilized monomers is unlikely. A second route of exposure is through skin absorption or ingestion during print handling because uncured monomers can leach onto the surface of the 3D-printed objects.*

This project gave the 2019 Greener Solutions student team the opportunity to contribute to a nascent but important public debate within a rapidly expanding industry in need of green chemistry innovation. The challenge of creating reactive cross-linking reactions without inherent chemical hazard highlighted the difficulties of incorporating green chemistry principles into the design of SLA resins, and the team provided several novel approaches for the AM industry to consider.

Who were involved

The 2019 Greener Solutions team was comprised of five students: Jacob Manheim participated as a Master of Public Health (MPH) student with experience in environmental consulting. Ladan Khandel contributed as an MPH student in industrial hygiene, with research experience focusing on acrylate exposures in occupational settings. Kyle Peerless, an MPH student with a focus in industrial hygiene, provided expertise in technical performance based on his previous research experience focused on mechanical properties of parts produced by AM (B.S. in mechanical engineering). Lauren Irie, a fourth-year undergraduate chemistry student, contributed her expertise in polymer chemistry research. Rachel Scholes, a sixth-year graduate student pursuing her Ph.D. in environmental engineering, contributed her background in photochemistry, chemical processes (B.S. in chemical engineering), and environmental contaminants. Tom McKeag represented the team's partner, the US EPA, presented the challenge and coordinated the team with other 3D printing experts, such as BCGC researcher Nicole Panditi, and chemist Justin Bours. Physician and School of Public Health faculty Megan Schwarzman, and USDA chemist Dr.William Hart-Cooper provided the essential, foundational instruction and problem-solving guidance in the course.

What was the solution

This team developed a framework for innovation in SLA resins, outlined in five strategies, and provided recommendations for further research and testing of the proposed alternative resins.

The Greener Solutions team first defined a scope of work based on their challenge statement, initial industry research, and conversations with their partners. They quickly identified the hazards associated with acrylate-based resins and were tasked with finding an improvement relative to PR48 as a baseline. They also learned why acrylates are prominent in existing resins. Through initial literature research and conversations with partners, they found that polymerization of acrylates is fast, has a well-known photoinitiated radical chain reaction mechanism, and that the acrylate monomers and oligomers can be varied to tune the mechanical properties of printed objects. They soon discovered that there would not be an off-the-shelf technology ready to replace existing resins, and they defined their scope to identify potential alternative chemistries for SLA printing, knowing they would have to rely on technologies that were not fully developed.

The team defined technical performance criteria based on initial research. They chose to focus on print speed, dimensional accuracy, and strength as metrics for technical performance. Because SLA is primarily used in settings where precision and speed are critical (e.g., hobbyist, research, and biomedical applications), the team decided to prioritize evaluating cure time and dimensional accuracy. Curing speed is critical to rapid, layer-by-layer prototyping, with existing SLA technology able to print each layer of resin in approximately 10 seconds. Viscosity was used as a proxy metric in some cases because it affects print speed by determining how quickly liquid resin flows around the moving build platform to form new layers. Both viscosity and curing speed also affect layer thickness, which in turn influences the dimensional accuracy of printed parts. Dimensional accuracy measures how closely the dimensions of a printed part match the design in the CAD file. Existing printers are able to create objects with dimensions within 0.10 mm of the expected value. Strength and structural integrity, measured with compressive or tensile strength tests, contribute to durable, long-lasting parts. Existing resins produce parts with strength metrics in the 1–10 MPa range. While other metrics were also identified, these comprised the primary technical performance criteria evaluated across each strategy (Table 5.9).

TABLE 5.9 Technical performance criteria for current and alternative formulations. From: Irie et al. (2019). 3D printing a safer future: Alternative resins for stereolithography, unpublished report as cited from Cheng, C., Dennis, A., Hill, L. A., Rainey, C., & Rodriguez, B. (2015). Autodesk and Stereolithography 3D Printing: Bio-inspired Resins for Better Human and Environmental.

Polymerization and Aesthetic	• 70-80% conversion or better • Smooth surface of printed product
Speed	• Material layer must cure in < 10 sec
Viscosity	• <0.5 Pa*s
Dimensional Accuracy	• +/- 0.10 mm
Layer Thickness	• 50-100um or smaller
Strength	• 1-10 MPa
Clarity	• Resin allows for light penetration
Cost and Commercial Availability	• $150/L or better

For health and environmental performance, the team developed criteria to prioritize solutions with significantly reduced hazards compared to PR48. The percent of unreacted reactive ingredients in the final print, irritation to skin and eyes, sensitization, and aquatic toxicity were of particular concern in the PR48 formula. In addition, other common health endpoints (e.g., carcinogenicity, general toxicity) were evaluated to ensure one hazard endpoint was not being swapped for another. Sustainability was another criterion, specifically whether the resin ingredients could be derived from renewable sources (Table 5.10).

The team then developed an initial list of potential replacements, taking inspiration from nature and from existing polymers. At first, they considered polymers synthesized in other contexts, such as polyolefins, which can be polymerized via a free radical mechanism similar to acrylates. However, polyolefins are not as reactive as (meth)acrylates because the radical intermediates are not stabilized by resonance (Konstantinov & Broadbelt, 2019), so the rate of polymerization could be insufficient. Other polymers used in industry today require high temperatures for polymerization, and they are generally not formed by covalent bond formation, but by thermal curing. Based on these findings, polyolefins made it on the initial alternatives list, but were not highlighted among the team's final alternatives, partly due to concerns regarding toxicity of the coordination catalysts required for polyolefin free radical polymerization (Grau, 2010).

In tandem, the team began to develop strategies using entirely different chemistries—getting away from free radical polymerization of monomers and oligomers. They looked for strategies that were bio-inspired, in the hope that these efforts would yield entirely new chemistries with lower hazard, and potentially with renewable feedstocks (Fig. 5.12).

First, the team looked for strong materials in nature, then searched for instances of their use in SLA printing or other 3D printing methods, then filled in data gaps where possible. For instance, strong natural materials are comprised of chitin and proteins in shellfish, and chitosan, derived from natural chitin, has been used in 3D printing (Hu & Gao, 2008). This finding became the basis for one strategy, named "functionalized biopolymers." This strategy relied on attaching methacrylate functional groups to biopolymers, making the biopolymers able to cross-link with each other. Having the methacrylate functional groups attached to large biomolecules reduced the hazard profile because the high molecular weight (>500 Da) would preclude skin penetration or inhalation. Functionalized chitosan, for instance, has been used to print hydrogels for biomedical applications, resulting in nontoxic products (Saikia et al., 2015). Along these lines, the team later found bovine serum albumin (BSA), a protein present in cow blood, had also been functionalized with methacrylates and used in an SLA printing context. Despite the high molecular weight of the BSA backbone, this strategy was able to achieve a fast print speed with high dimensional accuracy (Smith et al., 2019), giving the team confidence that other resins with high molecular weight components could also succeed.

TABLE 5.10 Health and environmental performance criteria.

	Goal	Target
Human Health	Low Oral Toxicity	LD50 > 5000 mg/kg
	Low Inhalation Toxicity	LD50 > 20 mg/L
	Non-irritating to Skin and Eyes	Negative results from in vitro irritation tests or animal tests
	Low Sensitization Potential	Negative results from animal tests
	Minimal Mutagenic and Carcinogenic Toxicity	Negative for chromosomal aberrations, Negative results from Ames test
	Minimal Reproductive or Developmental toxicity	No endocrine disruptors, Negative results from reproductive / developmental toxicity screening assay
	Minimal Unreacted Reactive Ingredients in Final Product	Less than 10%
Healthy Ecosystems	Biodegradable	Half life shorter than 40 days in freshwater
	Low Potential for Bioaccumulation	log Kow < 3
	Minimal Toxicity to Aquatic Life	96 hr LC50 > 100 mg/L
	Minimal Hazardous Waste Generated	No generation of hazardous waste
Sustainability	Chemicals Sustainably Sourced and Manufactured	Chemicals not derived from fossil fuels

From: Irie et al. (2019). *3D printing a safer future: Alternative resins for stereolithography,* unpublished report.

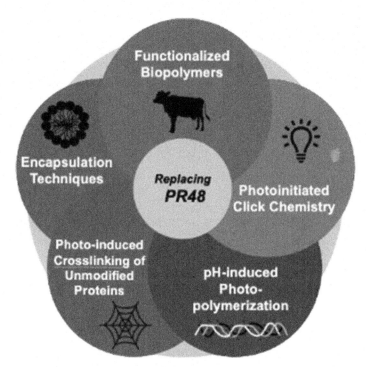

FIGURE 5.12 Overview of proposed strategies. *From: Irie et al. (2019).* 3D printing a safer future: Alternative resins for stereolithography, unpublished report.

The team was simultaneously considering other biopolymer-based strategies, focusing on strong, naturally occurring materials such as lignin and spider silk. For each material, they needed to establish how it would photocure and how quickly, and what material properties could be expected. The team presented their initial ideas on these materials and their research on the functionalized biopolymer strategy to their classmates and instructors as an intermediate deliverable, to receive feedback and suggestions.

From their presentation of the functionalized biopolymer strategy, the team decided it would be helpful to frame their strategies based on the reaction type (e.g., crosslinking of functionalized biopolymers) rather than the specific backbone (i.e., chitosan vs. lignin vs. spider silk). They realized that the bio-based polymers represented a theme running throughout their proposals, but that the reactions and curing processes differentiated each strategy by technical performance and hazard profile. They therefore reorganized their strategies based on the mechanism of polymerization, giving specific examples of materials that could be used with each. This change in framing made the team's findings in one strategy easier to apply to another by pointing out parallels between the different processes they had been researching.

For instance, the team built off the functionalized biopolymer strategy and research on existing polymers to suggest click chemistry-functionalized biopolymers. Click chemistry uses chemical functional groups that undergo specific, high-yield reactions to form covalent bonds. In particular, the click chemistry-based strategy developed by the 2019 Greener Solutions team used a known click chemistry reaction type, Diels–Alder cycloaddition, where functional groups attached to a biopolymer backbone selectively and efficiently cross-link without the need for a catalyst.

This strategy exhibited the value in using biopolymer backbones in resins, which reduced the hazard profile, and further aimed to completely remove acrylate functional groups. The alternative chemistries of this strategy still involved toxic components used to functionalize the biopolymers, particularly because they contained the reactive 1,4-unsaturated carbonyl feature present in acrylates (IARC, 2012; ToxNet, 2015), but this strategy presented another advantage: the proposed cross-linking reactions could be achieved without the use of a photoinitiator (Winkler et al., 2012). Though not the main focus of this team, the photoinitiator in PR48 also presented significant health hazards, so the removal of this component from the formula was an added health and environmental benefit (Table 5.11).

The 2019 Greener Solutions team had also been searching for a way to build off the work of the 2015 Greener Solutions team in suggesting pH-induced polymerization using photo base generators. One of the new strategies suggested improvements to the 2015 proposal that would use less hazardous materials.

The 2015 team had suggested cross-linking DOPA (dihydroxyphenylalanine), which is a developmental toxicant. The 2019 team combined the bio-inspired pH-induced polymerization strategy with an alternative protein, collagen, which has a safer hazard profile and had only recently been used in biomedical applications, including in 3D printing. They discovered the use of collagen-hydroxyapatite composites in bone scaffolding and inkjet printing, where the formation of cross-linked solids occurs by adjusting the pH of a liquid solution (Inzana et al., 2015; Lee et al., 2019; Wahl & Czernuszka, 2006). They then proposed to photoinitiate this process using the photo base generators that had been researched for the 2015 project.

Following the internal strategy presentation, the team further developed two additional strategies that thus far had not been developed specifically for SLA printing. Spider silk was identified as an inspiration for a biopolymer strategy, but instances of using unmodified proteins, like those comprising spider silk, in photopolymerization had not previously been identified. The team came across an analytical technique used to crosslink unmodified spider silk proteins and evidence that this technique could be used to print solid objects (Gill et al., 2018). This finding grew into their fourth strategy: cross-linking of unmodified proteins.

TABLE 5.11 Overall comparison of performance across product lifecycle.

| Strategy | Health Criteria | | Technical | Environmental |
	Formulation	Printing/Use	Printing/Use	Waste Disposal
PR48	3	3	1	3
Biobased Polymers	3	1	1	2
Click Chemistry	3	1	2	2
pH Induced	1	1	2	1
Spider Silk	2	2	2	2
Micelles	U	U	U	U

Legend			
(1) High Performance	(2) Medium Performance	(3) Low Performance	(U) Unknown

From: Irie et al. (2019). *3D printing a safer future: Alternative resins for stereolithography*, unpublished report.

In addition, the team wanted to explore the idea of reducing the amount of unreacted ingredients remaining on the final print, but in most cases had no evidence regarding this endpoint. They introduced a new strategy, called "encapsulation techniques," with the aim of reducing exposure to unreacted resin ingredients. This strategy uses micelles to contain reactive monomers and oligomers and release them upon irradiation. The micelles could also re-encapsulate unreacted components upon postprint exposure to visible light (Son et al., 2014). Though micelles had not been used in SLA printing previously, the team proposed that application-specific micelles could be developed and could greatly enhance the safety of printing. This strategy was the most unproven but was valuable in showcasing another novel approach to safety.

In December, 2019, the team presented these five strategies in a final report and presentation: (i) methacrylate-functionalized biopolymers, (ii) photo-initiated click chemistry, (iii) pH-induced photopolymerization, (iv) photo-induced cross-linking of unmodified proteins, and (v) encapsulation for release of reactive monomers. In the final analysis of these strategies, the team explicitly considered life cycle impacts of health and environmental endpoints. This framing highlighted where the improvement was, and for whom risk would be reduced. For instance, in the functionalized biopolymer strategies, risk remained for resin formulators who would be responsible for functionalizing the biopolymer backbones with reactive ingredients, but the risk to people using SLA printers was reduced.

What was innovative about the solution

Each of the 2019 team's strategies relied on innovative thinking to address the seemingly inherent contradiction between the reactivity required for photopolymerization and the aim of low toxicity. Existing SLA technology relies almost exclusively on low molecular weight, reactive ingredients. The molecules used as monomers and oligomers in SLA resins present a hazard because their reactivity toward polymerization comes with reactivity toward biomolecules. The team knew they needed cross-linking reactions to occur, but they also needed to protect human health. Two of their strategies (the pH-induced strategy and cross-linking of unmodified proteins) achieved high reactivity without health hazards by applying technologies from other fields in an SLA context. Their other strategies focused on reducing the hazard to users by decreasing exposure —either by increasing the molecular weight of the reactive components to decrease dermal and inhalation exposure, or by encapsulating the reactive components and thereby physically separating them from the user.

For some strategies, technologies were applied from other industries to SLA printing. For instance, the pH-induced photopolymerization strategy used the demonstrated formation of strong collagen-based materials via pH change and attached a photoinitiated mechanism using photo base generators. Strong collagen-hydroxyapatite composite materials were used in bone scaffolding and had been inkjet printed by precise pH control (Lee et al., 2019), but these materials had not been previously demonstrated in an SLA printer. The proposal to pair these materials with photo base generators discussed by the 2015 Greener Solutions team represented the application of an existing technology with a new photoinitiation method.

Another example of innovation was the use of postprint processing steps to improve technical performance. Particularly in the biopolymer-based strategies, significant literature existed on creating bio-based hydrogels, but far fewer applications required the stronger, more durable materials needed for SLA-printed objects. Postprint curing was identified as a means to transform BSA-based hydrogels into bioplastics that met the technical performance strength criteria (Smith et al., 2019). Additives such as bioactive glass or collagen could also be used to increase the strength of printed parts (Jarosz et al., 2019). Combined with the wealth of existing research on hydrogels, these modifications could be applied across a range of materials and represented the formation of new, much-improved substances.

What was the impact

From the 2019 team's work, industry and research partners received a roadmap and avenues for further consideration. In particular, the efforts provide avenues for further research and development by industry partners. The ideas presented are novel and can be used to inspire new techniques in the SLA printing. It is still unclear what broader impacts this effort will have, but these proposals have widened the range of possibilities within the AM industry and hopefully will advance the pursuit of safer alternatives in SLA printing.

For the first time ever, Greener Solutions graduate student teams, mentored by BCGC as part of this project, proposed a SLA polymerization system that avoided the use of acrylate or methacrylate cross-linkers and instead employed a collagen-ribose-hydroxyapatite resin formula with light and pH change-activated polymerization system.

The project resulted in proposed solutions that could potentially provide great value to the professional SLA community by highlighting opportunities for safer resins. The team went beyond replacing specific components to fully redesign SLA resins with renewable feedstocks, reduced hazard, and new strategies like additives and postprint processing. These solutions highlighted new places in the SLA printing system that could be manipulated to improve safety. While the proposals developed by the 2019 Greener Solutions team are not directly useful for setting regulations, they do give regulators and policymakers a sense of the opportunities for safer alternatives in SLA printing and an overview of where additional research and development could be beneficial.

TABLE 5.12 Development stages of proposed strategies.

	Preliminary Research	Proof-of-concept	Assess areas for improvement	Prototyping
Biopolymers				Consider post-print curing for enhanced strength
Click chemistry			Test modifications to improve print speed	
pH induced		Assess alternative photo base generators, test print with hydroxyapatite		
Spider silk		Test layer-by-layer printing Consider post-modification. Ru alternatives		
Micelles	Find research partners to develop & demonstrate technology/			

From: Irie et al. (2019). *3D printing a safer future: Alternative resins for stereolithography,* unpublished report.

Next steps: feasibility, data gaps, and requirements for progress in implementation

The team recommended next steps in technology development ranging from preliminary research to prototyping for the proposed strategies. Further optimization of postprint curing steps and additives that improve mechanical properties could improve multiple strategies. In addition, the print speed for click chemistry-functionalized biopolymers and pH-induced polymerization requires further testing and optimization. These resin formulas will require further tinkering of component ratios and functionalization density. The selection of a less hazardous photo base generator should also be prioritized. Implementation overall will require further research and development, which would be aided by investment from industry actors interested in implementing these techniques into new sustainable and healthy resins (Table 5.12).

Overall summary of BCGC work in additive manufacturing

To date, BCGC work in the AM sector has comprised six phases, including professional consulting, chemical alternatives investigations via the Greener Solutions graduate course, the creation of assessment strategies and tools, and engagement with industry and researchers This work has resulted in several products: an assessment of relative hazard of SLA resin formulations and recommendations for researching alternatives; proposed nature-based strategies for replacement of current resin formulations, a "biofriendly" database of material choices; a material assessment scorecard combining several systems of assessment at different life cycle stages; and a user guide for the assessment tool.

References

The American Society for Testing and Materials (ASTM) group "ASTM F42—Additive Manufacturing (2012) Standard Terminology for Additive Manufacturing Technologies. https://www.astm.org/Standards/F2792.htm.

Bours, J., Adzima, B., Gladwin, S., Cabral, J., & Mau, S. (2017). Addressing hazardous implications of additive manufacturing: Complementing life cycle assessment with a framework for evaluating direct human health and environmental impacts. *Journal of Industrial Ecology, 21*(S1), S25—S36.

California Department of Toxic Substances Laws (22 CCR §66261.20-261.24). https://govt. westlaw.com/calregs/Browse/Home/California/CaliforniaCodeofRegulations?guid=IA16A08 E0D4BA11DE8879F88E8B0DAAAE&originationContext=documenttoc&transitionType= Default&contextData=(sc.Default.

Chan, K., & O'Brien, P. J. (2008). Structure—activity relationships for hepatocyte toxicity and electrophilic reactivity of α, β-unsaturated esters, acrylates and methacrylates. *Journal of Applied Toxicology, 28*(8), 1004—1015.

Chen, C., Dennis, A., Hill, L. A., Rainey, C., & Rodriquez, B. (2015). *Autodesk and stereolithography 3D printing: Bio-inspired resins for better human and environmental health.* Unpublished report available at https://bcgctest.files.wordpress.com/2017/04/final-reportautodesk_greenersolutions_2015.pdf.

Grau, E. (2010). *Polymerization of ethylene: From free radical homopolymerization to hybrid radical/catalytic copolymerization. Polymers thesis. L'Universite Claude Bernard — Lyon I.*

Gill, E. L., Li, X., Birch, M. A., & Huang, Y. Y. S. (2018). Multi-length scale bioprinting towards simulating microenvironmental cues. *Bio-design and Manufacturing, 1*(2), 77−88.

Hu, X., & Gao, C. (2008). Photoinitiating polymerization to prepare biocompatible chitosan hydrogels. *Journal of Applied Polymer Science., 110*(2), 1059−1067.

Inzana, J. A., Olvera, D., Fuller, S. M., Kelly, J. P., Graeve, O. A., Schwartz, E. M., Kates, S. L., & Awad, H. A. (2014). 3D printing of composite calcium phosphate and collagen scaffolds for bone regeneration. *Biomaterials, 35*(13), 4026−4034.

Jarosz, T., Gebka, K., & Stolarcyk, A. (2019). Recent advances in conjugated graft copolymers: Approaches and applications. *Molecules, 24*(16), 3009.

Konstantinov, I. A., & Broadbelt, L. J. (2019). A quantum mechanical approach for accurate rate parameters of free-radical polymerization reactions. In *Computational Quantum Chemistry* (pp. 17−46). Elsevier.

Lee, A., Hudson, A. R., Shiwarski, D. J., Tashman, J. W., Hinton, T. J., Yerneni, S., Bliley, J. M., Campbell, P. G., & Feinberg, A. W. (2019). 3D bioprinting of collagen to rebuild components of the human heart. *Science, 365*(6452), 482−487.

McKeag, T. A. (2010). Shaping the future of additive manufacturing: Twelve themes from bio-inspired design and green chemistry. *Handbook of Green Chemistry: Online*, 241−262.

Mulvihill, M., Bours, J., & McKeag, T. (2015). *Toward proactive materials substitution in the resin formulations for the Autodesk Ember SLA printer: A biomimetic approach.* Unpublished paper.

Saikia, C., Gogoi, P., & Maji, T. K. (2015). Chitosan: A promising biopolymer in drug delivery applications. *Journal of Molecular Genetic Medicine, 4*(006).

Smith, P. T., Narupai, B., Tsui, J. H., Millik, S. C., Shafranek, R. T., Kim, D. H., & Nelson, A. (2019). Additive manufacturing of bovine serum albumin-based hydrogels and bioplastics. *ChemRxiv [Preprint]*, 9758876.

Son, G. M., Kim, H. Y., Ryu, J. H., Chu, C. W., Kang, D. H., Park, S. B., & Jeong, Y. I. (2014). Self-assembled polymeric micelles based on hyaluronic acid-g-poly (D, L-lactide-co-glycolide) copolymer for tumor targeting. *International Journal of Molecular Sciences, 15*(9), 16057−16068.

ToxNet. (2015). *Benzophenone.* Available at: https://toxnet.nlm.nih.gov/cgi-bin/sis/search/a?dbs+hsdb:@term+@DOCNO+6809.

US EPA Laws 40 CFR 261.21-261.24: www.gpo.gov/fdsys/pkg/CFR-2015-title40-vol26/pdf/CFR-2015-title40-vol26-chapIsubchapI.pdf.

Wahl, D. A., & Czernuszka, J. T. (2006). Collagen-hydroxyapatite composites for hard tissue repair. *European Cells and Materials, 11*, 43−56.

Winkler, M., Mueller, J. O., Oehlenschlaeger, K. K., de Espinosa, L. M., Meier, M. A. R., & Barner-Kowollik, C. (2012). *Highly orthogonal functionalization of ADMET polymers via photo-induced diels-alder reactions, 45 (12)* (pp. 5012−5019).

Further reading

Bours, J., Adzima, B., Gladwin, S., Cabral, J., & Mau, S. (2017). Addressing hazardous implications of additive manufacturing: Complementing life cycle assessment with a framework for evaluating direct human health and environmental impacts. *Journal of Industrial Ecology, 21*(S1), S25−S36.

Braungart, M., McDonough, W., Kälin, A., & Bollinger, A. (2012). *Cradle-to-cradle design: Creating healthy emissions—a strategy for eco-effective product and system design* (pp. 247—271) (Birkhäuser).

Campbell, I., Diegel, O., Kowen, J., & Wohlers, T. (2020). *Wohlers report 2020: 3D printing and additive manufacturing state of the industry: Annual worldwide progress report.* Wohlers Associates. Available at https://wohlersassociates.com/2020report.htm.

Clean Production Action. (2013). *GreenScreen for safer chemicals chemical hazard assessment procedure.* Available at: http://www.greenscreenchemicals.org/static/ee_images/uploads/resources/GreenScreenv1 -2_Guidance_Assessment_Procedure_FINAL_2013_9_18.pdf.

Crivello, J. V., & Reichmanis, E. (2014). Photopolymer materials and processes for advanced technologies. *Chemistry of Materials, 26*(1), 533—548.

Faludi, J., Bayley, C., Bhogal, S., & Iribarne, M. (2015). Comparing environmental impacts of additive manufacturing vs traditional machining via life-cycle assessment. *Rapid Prototyping Journal.*

Faludi, J., Hoang, T., Gorman, P., & Mulvihill, M. (2016). Aiding alternatives assessment with an uncertainty-tolerant hazard scoring method. *Journal of Environmental Management, 182*, 111—125.

Fernandez, J. G., & Ingber, D. E. (2014). Manufacturing of large-scale functional objects using biodegradable chitosan bioplastic. *Macromolecular Materials and Engineering, 299*(8), 932—938.

Heller, C., Schwentenwein, M., Varga, F., Liska, R., & Stampfl, J. (2009). Biocompatible and biodegradable photopolymers for microstereolithography. *Proceedings of LAMP.*

International Agency for Research on Cancer. (2012). *IARC Monographs: Benzophenone.* Available at https://monographs.iarc.fr/wp-content/uploads/2018/06/mono101-007.pdf.

Jacobs, P. F. (1992). *Rapid prototyping & manufacturing: Fundamentals of stereolithography.* Society of Manufacturing Engineers.

Lang, N., Pereira, M. J., Lee, Y., Friehs, I., Vasilyev, N. V., Feins, E. N., ... Pedro, J. (2014). A blood-resistant surgical glue for minimally invasive repair of vessels and heart defects. *Science Translational Medicine, 6*(218), 218ra6—218ra6.

McKeag, T. A. (2010). Shaping the future of additive manufacturing: Twelve themes from bio-inspired design and green chemistry. *Handbook of Green Chemistry: Online*, 241—262.

Monzón, M. D., Ortega, Z., Martínez, A., & Ortega, F. (2015). Standardization in additive manufacturing: Activities carried out by international organizations and projects. *The International Journal of Advanced Manufacturing Technology, 76*(5—8), 1111—1121.

Rickett, T. A., Amoozgar, Z., Tuchek, C. A., Park, J., Yeo, Y., & Shi, R. (2011). Rapidly photo-cross-linkable chitosan hydrogel for peripheral neurosurgeries. *Biomacromolecules, 12*(1), 57—65.

Scholes, R., Irie, L., Manheim, J., Peerless, K., & Khandel, L. (2019). *3D printing a safer future: Alternative resins for stereolithography.* Unpublished report available at https://bcgctest.files. wordpress.com/2020/01/slachallenge_5052931_76365293_sla_final_report.pdf.

http://www.chemspider.com/.

http://www.nasonline.org/about-nas/history/archives/milestones-in-NAS-history/organization-of-the-nrc.html.

https://bcgc.berkeley.edu/greener-solutions-2019/.

https://dtsc.ca.gov.

https://ecology.wa.gov.

https://home.dartmouth.edu/.

https://investors.autodesk.com/news-releases/news-release-details/autodesk-announces-100-million-spark-investment-fund-worlds.

https://knowledge.autodesk.com/support/netfabb/learn-explore/caas/blog/blogs.autodesk.com/netfabb/2015/11/18/towards-sustainable-biofriendly-materials-for-additive-manufacturing-part-1-of-3.html?us_oa=akn-us&us_si=36141c09-2751-4cc8-b7ed-1f4625376fc6&us_st=3d%20printing%20%2B%20berkeley.

https://knowledge.autodesk.com/support/netfabb/learn-explore/caas/blog/blogs.autodesk.com/netfabb/2016/01/21/how-nature-makes-things-relevant-bio-inspired-approaches.html?us_oa=akn-us&us_si=b9d0b679-b119-48d5-9a8f-71d780314129&us_st=3d%20printing%20%2B%20berkeley.

https://knowledge.autodesk.com/support/netfabb/learn-explore/caas/blog/blogs.autodesk.com/netfabb/2016/01/22/towards-sustainable-biofriendly-materials-for-additive-manufacturing-part-2-of-3.html?us_oa=akn-us&us_si=6e1dd507-e789-46d5-8e3a-af9005e10135&us_st=3d%20printing%20%2B%20berkeley.

https://knowledge.autodesk.com/support/netfabb/learn-explore/caas/blog/blogs.autodesk.com/netfabb/2016/05/23/towards-sustainable-biofriendly-materials-for-additive-manufacturing-part-3-of-3.html?us_oa=akn-us&us_si=36141c09-2751-4cc8-b7ed-1f4625376fc6&us_st=3d%20printing%20%2B%20berkeley.

https://methodhome.com.

https://pubmed.ncbi.nlm.nih.gov/.

https://theic2.org/#gsc.tab=0.

https://unece.org/about-ghs.

https://www.astm.org/Standards/additive-manufacturing-technology-standards.html.

https://www.autodesk.com.

https://www.c2ccertified.org/about/about.

https://www.c2ccertified.org/resources/detail/exposure-assessment-methodology.

https://www.chemforward.org/.

https://www.cpspolymers.com/about.

https://www.dailycal.org/2020/09/30/environmental-protection-agency-awards-grant-to-uc-berkeley-program-for-pollution-prevention/.

https://www.epa.gov/p2/grant-programs-pollution-prevention.

https://www.greenbiz.com/article/autodesk-and-pursuit-bio-inspired-3d-printing.

https://www.iso.org/standard/75142.html.

https://www.nist.gov/.

https://www.northwestgreenchemistry.org/.

https://www.northwestgreenchemistry.org/news/northwest-green-chemistry-initiates-study-on-sustainable-additive-manufacturing-3d-printing-materials.

https://www.osha.gov/laws-regs/standardinterpretations/1997-01-30-1.

https://www.osha.gov/sites/default/files/publications/OSHA3514.pdf.

https://www.radtech.org.

https://www.sigmaaldrich.com/united-states.html?gclid=Cj0KCQjw2NyFBhDoAR-IsAMtHtZ5n1LUlqIMLxB5bZB_pnZ5SkgNqIIuaKhNml2OMSPDErBi-TAmkteXcaAmczEALw_wcB.

https://www.tudelft.nl/en/.

Chapter 6

Green Chemistry case study on alternative energy: making the case against coal in Kosovo

What was the challenge

The World Bank established a new policy in 2013, that restricted lending capital for the development of new coal-fired power plants, except in cases wherein there was no financially reasonable alternative. One such exception to this new policy was a proposal for a new coal-fired plant to replace an old coal-based facility in the Republic of Kosovo, a country that has relied on coal for 98% of its electricity generation. The challenge taken up by BCGC-sponsored researchers was to model a renewable energy plan for Kosovo, including an alternatives assessment to demonstrate that there were more sustainable alternatives to building a new coal power plant. Persuading the World Bank not to invest in new coal plants would send a powerful message to nations seeking external funding for new energy production methods, and could have global consequences for cleaner energy investment around the world (Fig. 6.1).

The decommissioning of the Kosovo A coal-fired power plant in 2017, was the impetus for the international lending community to consider loans to support a new 600 MW facility to replace it. Kosovo's aging coal-fired power stations had been fired with the locally available lignite coal for decades. Lignite is a sedimentary rock; the intermediate stage between peat and anthracite (hard or "black" coal) and was formed relatively recently in geologic time, 50 million years ago. Lignite, or so-called "brown coal" is the poorest grade of coal for producing energy, and, as opposed to the highest grade, anthracite, is of low density, contains a low amount of carbon (60%−70%) and high latent moisture content (40%−60%). It, therefore, contains typically about 16.5 MJ/kg of energy, while anthracite is usually in the range of 16.5−32.5 MJ/kg. As a result of this low potential energy, more brown coal needs to be burned to produce the same amount of energy. More burning means more emissions from the stacks of the power plants using coal.

Brown coal is typically used for electricity generating power plants, as in Kosovo, and because of the poor grade is not generally shipped any great

Green Chemistry in Practice. https://doi.org/10.1016/B978-0-12-819674-8.00006-0

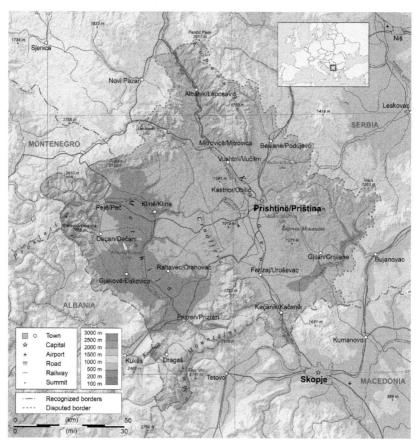

FIGURE 6.1 Map of Kosovo. *Source: https://upload.wikimedia.org/wikipedia/commons/d/db/ Kosovo_map-en1.svg.*

distance, but instead, power plants are located near the source. The coal is extracted by surface mining, also a significant health and environmental concern because it occurs above ground, destroys landscapes, and creates its own significant air pollution. All of these conditions existed in Kosovo. Approximately 8.8 million metric tons of lignite coal had been produced in Kosovo in 2016, and Kosovo was not alone in the region. In the Western Balkans countries (Albania, Bosnia & Herzegovina, Kosovo, Macedonia, Montenegro, and Serbia) all of the existing coal power plants run on lignite.

The challenge was to model the baseline coal use's effect on human and environmental health with meaningful and transferrable metrics, develop a safer renewable energy alternative, and compare the energy performance of both.

Why this project was important

Combusted lignite coal is the most harmful to health because of its composition and low energy potential and produces the highest carbon dioxide emissions per megawatt of energy produced. All burned coal will release particulate matter (PM), sulfur dioxide (SO_2), nitrogen oxides (NOx), and other pollutants into the air, and these emissions can travel thousands of kilometers. Lignite is likely to release more sulfur and ash, and, due to its makeup and the need to burn more of it, more PM. Burning coal for electricity generation, as was done in Kosovo, is one of the biggest sources of industrial CO_2 emissions in the world, and a major contributor to climate change. More than 6600 coal-fired plants serve an estimated five billion people globally and contribute 46% of CO_2 emissions. Man-made carbon dioxide production is a major factor in the destruction of the ozone layer protecting the earth's atmosphere and the consequent climate change we are now undergoing. Coal-fired plants, particularly those using brown or soft coal, are, therefore, both a long-term and short-term threat to health.

Gases and particulate matter (PM) from coal combustion are harmful for a number of reasons. PM is the term for tiny particles found in the air. It is formed in the atmosphere because of chemical reactions between pollutants. These particles include dust, dirt, soot, smoke, and liquid droplets. PM is categorized by size and therefore relative intrusion into the lungs, PM-10 (10 μm in diameter and below) being coarse, and PM-2.5 being fine and of much greater concern. Air emissions from coal-fired power plants can contain the following substances of concern. Please note we have not differentiated between acute (short term) and chronic (long term) exposure, exposure pathways, or vulnerability of population in the items in this list:

Cadmium: known carcinogen (US Department of Health and Human Services (DHHS)), severe lung damage, stomach, and respiratory irritation, fragile bones, and possible kidney disease.

Chromium: Chromium VI: known carcinogen (US DHHS, EPA; WHO IARC), lung cancer, stomach ulcers, respiratory irritation, skin irritation, and sensitization.

Dioxins/furans (include PCBs and Tetrachlorodibenzo-p-Dioxin or TCDD): 2,3,7,8-TCDD: human carcinogen (WHO IARC), chloracne, a severe skin disease, possible liver damage, endocrine disruption. Immune system disruption and birth defects.

Formaldehyde: known human carcinogen (US DHHS), nasal and eye irritation, bronchitis, increased risk of asthma/allergy, liver and kidney damage.

Hydrogen chloride: this is the gaseous state of hydrochloric acid (HCl): corrosive to the eyes, skin, and mucous membranes, may cause coughing, hoarseness, inflammation, and ulceration of the respiratory tract, chest pain, and pulmonary edema in humans, corrosion of the mucous membranes, esophagus, and stomach, severe skin burns, ulceration, pulmonary irritation, lesions of the upper respiratory tract, and laryngeal and pulmonary edema, gastritis, chronic bronchitis, dermatitis, and photosensitization.

Hydrogen fluoride: severe respiratory damage including severe irritation and lung edema. Severe eye irritation and skin burns, skeletal fluorosis, a bone disease.

Lead: probable human carcinogen (US DHHS, EPA; WHO IARC) nervous system damage, anemia, kidney damage, blood pressure increase, brain damage, reproductive, death.

Nickel: nickel subsulfide: human carcinogen (US EPA), nickel carbonyl: probable human carcinogen (US EPA); nickel compounds: human carcinogens (WHO IARC), chronic bronchitis, reduced lung function, and cancer of the lung and nasal sinus, lung and kidney damage, dermatitis, pulmonary thrombosis, and renal edema, possible reproductive and developmental impairment.

Particulate matter: probable human carcinogen when diesel exhaust (IARC), heart attack, irregular heartbeat, aggravated asthma, respiratory irritation, premature death in those with heart or lung disease.

Polycyclic aromatic hydrocarbons (PAHs): are reasonably expected to be human carcinogens (US DHHS), and possible reproductive and immune system damage.

Sulfur dioxide: potentially fatal above 100 ppm exposure, lung damage, respiratory irritation.

Volatile organic compounds (VOCs): some are known carcinogens (IARC), eye and respiratory irritation, liver, kidney, and central nervous system damage.

Radioactive contaminants: Burning coal also releases uranium, thorium, ruthenium, and other radioactive isotopes in concentrated form. Even at low levels, these isotopes can accumulate in the human body and form life-long deposits in bones and teeth.

Hazardous air pollution released by coal-fired power plants can cause a wide range of health effects. Exposure to coal power plant pollution can damage the brain, eyes, skin, and breathing passages. It can affect the kidneys, lungs, and nervous and respiratory systems. Health conditions from exposure to the pollutants from coal-fired power plants have been well documented. They include premature deaths, cardiovascular diseases, lung cancer, low birth weights, higher risk of developmental and behavioral disorders in infants and children, and higher infant mortality.

In addition to air emissions, both coal mining and combustion create solid and liquid waste that is of concern. Mining creates a leftover mixture of coal, soil, and rock, and power plants create liquid coal waste from the washing process and ash from burning. While the fly ash from burning can be used as an additive to cement, it can also be a large source of air, water, and soil pollution if improperly stored or released into the environment. Fly ash can contain silica and the toxic metals cadmium, copper, chromium, nickel, lead, mercury, titanium, as well as arsenic and selenium.

This technical challenge was further complicated by geopolitical concerns within a region still impacted by a tragic and painful transition. The 1998—99 Kosovo war erupted when ethnic Albanians sought to secede from the Federal Republic of Yusgoslavia, (the Republics of Serbia and Montenegro), which was dominated in the government and military by ethnic Serbs. In a conflict rife with retributions inflicted on civilians and combatants alike, approximately 13,500 people were killed, and between 1.2 and 1.45 million Kosovo Albanians and approximately 200,000 ethnic Serbs were displaced.

Within this context, researchers had these three additional considerations:

1. In contrast to Kosovo's legacy and proposed coal plants, which would be part of a centralized power grid, the renewable energy solutions proposed by BCGC researchers would likely lead to decentralized energy production, so the researchers needed to provide a compelling model for the logistical and technical feasibility of decentralized energy production.
2. The low cost and abundance of lignite coal in Kosovo made it an attractive energy source, so the case needed to be made that the human and environmental health costs of lignite coal exceeded the economic benefits of continuing to use it.
3. Regional geopolitics complicated the development of a decentralized energy model, necessitating the development of a national energy model that matched the interests and constraints of Kosovars.

Who were involved

The Berkeley Center for Green Chemistry provided National Science Foundation (NSF) IGERT funding to BCGC SAGE fellow and Energy and Resources Group graduate student, Noah Kittner, to research energy policy and travel to Kosovo in the summer of 2014, so that he could work with the local NGO Kosovo Civil Society Consortium for Sustainable Development (KOSID), the European Energy Commission, and the University of Pristina to develop a model for renewable energy in Kosovo.

Kittner's energy systems modeling built on the previous work of his graduate advisor, Dr. Daniel M. Kammen (Director of the Renewable & Appropriate Energy Laboratory [RAEL] at UC Berkeley) that had connected renewable energy to large-scale geopolitical, economic, and human health impacts. With Kammen's help, Kittner acquired lignite coal samples while in Kosovo and brought them back to Berkeley for a more thorough health impact analysis.

At the Center for Green Chemistry, former center director Dr. Martin Mulvihill and researcher Dr. Heather Buckley suggested analyzing the coal samples using mass spectrometry to determine trace metal content, which would provide information about potential air pollution and public health impacts of lignite coal energy systems. Berkeley undergraduate researchers Raj Fadadu and Zac Mathe, assisted with the analysis of the coal samples. BCGC associate directors and professors from the Berkeley School of Public Health Dr. Tom McKone and Megan Schwarzman MD guided efforts to model human health impacts due to the combustion of lignite coal.

What was the solution

Led by Noah Kittner, the research team addressed each of the subchallenges in turn:

1. A new model for renewable energy systems

The researchers developed and published an open-source model for renewable energy systems that combined production costs, distribution impacts, and health impacts of different sustainable energy strategies. Over the course of several peer-reviewed publications, Kittner made the case for small-scale decentralized renewable energy sources (solar panels, wind power, mini-hydropower systems) as a viable alternative to centralized, carbon-heavy energy strategies. In 2016, Kittner collaborated with researchers at KOSID to publish a complete analytical framework for conducting energy policy alternative assessments. This framework was made freely available to foster informed public debate about renewable energy markets and technical performance and would eventually be used by the World Bank to guide energy investment policy.

This 2016, analytical framework was designed to provide "… a quick and low-budget first-cut analysis for comparison of technical and financial feasibility and cost-effectiveness of alternatives." The World Bank had determined that cost-effectiveness was to include capital and operational cost and financial benefits of the options considered over the life of the project.

The team noted that the current supply and demand situation in Kosovo had some inherent problems. Frequent blackouts and unmet demand were chronic. Although more than 95% of electric power generation came from the combustion of local lignite coal, Kosovo had had to import electricity, and this had serious energy supply security implications, as most of these imports (54%) would have to come from Serbia, its regional antagonist. Electricity generation had fallen off, and demand was artificially constrained by infrastructure limitations. Moreover, Kittner's team demonstrated additional unmet demand using UN FAO estimates of biomass fuels consumed for household heating in rural homes, itself an additional household air pollution health concern.

The Kosovar government had planned to generate 25% of its energy from renewables by 2020, was using feed-in tariffs to support this policy, and had enacted a law requiring first preference to the purchase of domestic production over trading arrangements. Although Kosovo had a nascent and small solar industry, as well as wind and small river-run hydropower potential, it was unclear if the country could meet its own goal or the higher standard required by the European Union, which Kosovo aspired to join. The team made the case that meeting EU energy and climate targets could expedite the process of gaining membership. The European Commission had enacted stricter greenhouse gas emission reduction targets along with increased energy efficiency, renewable generation goals, and plans for expanding regional interconnections. If Kosovo decided to adhere to the 2030 EU climate and energy framework outlined in the 2015 Industrial Emissions Directive, it would have to reduce its greenhouse gas production by 40% from 1990 levels, meaning that it would have to raise its renewable energy generation to 27% and improve energy efficiency by 27%.

The team calculated levelized cost of energy (LCOE) of electricity generation for several alternatives against the baseline coal-fired power plant by creating an annual generation spreadsheet model. This was a scenario approach that included policy parameters, target capacities, estimates of potential generation, and life cycle costs. The baseline case of "business as usual" of World Bank approval of loans for a Kosovo C coal plant included a shadow price of CO_2 emissions when using coal. Ultimately, the research team estimated the construction of Kosovo C could add up to 11.5 million tons of CO_2 per year, adding an additional amortized cost of 330 million euros for the plant.

The study compared seven scenarios, including solar power, the EU 2030 aggressive energy efficiency measures, regional transmission networks, and natural gas introduction.

The energy production options were annualized over 12 years, costs of renewables based on estimated capacities, and assumptions made about the timing of development and relative market costs within scenarios. Grid consumption data from the Ministry of Economic Development and population growth projections informed the comparison (Table 6.1).

The results indicated a wide range of energy alternatives that could meet the energy needs of the Kosovars at less than the cost of a coal-fired plant. It highlighted that a multi-faceted approach could be devised to better advantage, both in the short and long terms. For example, curtailing peak demand by energy efficiency, combined with solar power that is supplemented with natural gas to cover cloudy days. Each of the scenarios was shown to provide electricity until 2025 at a cost of less than 1.5−1.8 billion euros, significantly less than the estimated cost of 1.9−2.2 billion euros for the coal-based scheme. The authors stressed that a coal-centric future was not an economic or political necessity, and that the diverse set of decentralized, low-carbon pathways could be flexibly planned and adjusted over time, but that it would take forethought, discussion, and planned action (Table 6.2).

TABLE 6.1 Electricity supply options, Kosovo.

Scenario	Name	Notes
1	Base case (coal)	TPP C built in 2017, 2–300 MW turbines
2	Solar prices reduce to SunShot levels	Solar at €0.9 W⁻¹ by 2020; €30 ton⁻¹ of CO₂
3	Euro 2030 path: aggressive energy efficiency measures (27% increase), 27% CO₂ reduction, 27% renewable consumption along with expanded open regional market via a power exchange	1 kWh energy avoided displaces 1 kWh coal-fired generation
4	Regional transmission network allows for expanded electricity imports	Solar at €1 W⁻¹ by 2020 and imports dominate from Hungarian Power exchange
5	Introduction of natural gas via TAP by 2018 with aggressive energy efficiency measures	Solar at €1 W⁻¹ by 2020
6	Including a carbon shadow price	€30 ton⁻¹ of CO₂ added to cost of coal generation
7	Including storage cost for solar at high deployment levels	Solar at €1 W⁻¹ by 2020 and storage is €200 kWh⁻¹
	No natural gas, extra transmission for Albania–Kosovar joint projects	

From Kittner, N., Gheewala, & Kammen, D. M., (2016). *Environmental Research Letters*, 11, 104013, Table 3, "A selection of the multiple pathways examined in this paper that economically and reliably meet Kosovo's projected future electricity demand".

2. A more comprehensive assessment of health impacts

An important element of the energy production policy model that Kittner and others developed was the inclusion of health impacts as a cost component of energy markets; this allowed the researchers to more fully address the human and environmental health concerns raised by the extensive use of lignite coal as a primary energy source. Though previous studies had addressed the particulate air pollution generated from coal combustion, there was no data on the effects of exposure to trace metals released from lignite coal plants. This was a significant gap in the health cost calculation framework.

This part of the investigation had three interconnected parts: the first was a bench-top assay of the heavy metal components found within local lignite coal samples. The cataloging of metals contained in the samples allowed the combined team, in the second part, to develop a novel metric: mg/kWh, a measure of the mass of trace metal released per kilowatt hour of electricity generated. The third part was modeling of mortality and morbidity estimated from projected emissions shown to contain these elements. This allowed the team to present quantified human and environmental health costs for the energy production scenarios originally proposed by Kittner et al. This systems approach to a complex problem combined three analyses to evaluate the environmental, economic, and human health costs of energy production alternatives in Kosovo (Fig. 6.2).

The researchers found that trace metals are rarely accounted for in PM emission indices, despite the fact that they are clearly shown in geologic assays of coal. They believed that an investigation of these components and a

TABLE 6.2 Total cost estimates of electricity supply options.

Scenario	Name	Notes	Estimated cost[a]	Average LCOE	Figure
1	Base case (coal)	"New Kosovo" b built in 2017; 2–300 MW turbines	€1.96 billion EUR (€2.29 billion EUR with €30/ton⁻¹ CO₂ price; €1.86 billion EUR at 500 MW)	€204 MWh⁻¹ (€184 MWh⁻¹–€224 MWh⁻¹)	Figure 1; appendix table A1
2	Solar prices reduce to SunShot levels	Solar at 60.9 W⁻¹ by 2020; €30 ton⁻¹ of CO₂	€1.67 billion EUR	€165 MWh⁻¹ (€156 MWh⁻¹–€174 MWh⁻¹)	Figure 2; appendix table A2
3	Euro 2030 path: aggressive energy efficiency measures (27% increase, 27% CO₂ reduction, 27% renewable consumption along with expanded open regional market via a power exchange	1 kWh energy avoided displaces 1 kWh coal-fired generation	€1.57 billion EUR	€160 MWh⁻¹ (€150 MWh⁻¹–€170 MWh⁻¹)	Figure 3; appendix table A3
4	Regional transmission network allows for expanded electricity imports	Solar at €1 W⁻¹ by 2020 and imports dominate from Hungarian Power exchange	€1.76 billion EUR	€167 MWh⁻¹ (€162 MWh⁻¹–€172 MWh⁻¹)	Figure 4; appendix table A4
5	Introduction of natural gas via TAP by 2018 with aggressive energy efficiency measures	Solar at €1 W⁻¹ by 2020	€1.55 billion EUR	€155 MWh⁻¹ (€141 MWh⁻¹–€169 MWh⁻¹)	Figure 5; appendix table A5
6	Including a carbon shadow price	€30/ton⁻¹ of CO₂, added to cost of coal generation	€1.78 billion EUR	€169 MWh⁻¹ (€160 MWh⁻¹–€178 MWh⁻¹)	Figure 6; appendix table A6
7	Including storage cost for solar at high deployment levels	Solar at €1 W⁻¹ by 2020 and storage penalty at €200/kWh⁻¹, representing 10% of system generation costs	€1.57 billion EUR	€157 MWh⁻¹ (€150 MWh⁻¹–€164 MWh⁻¹)	Not pictured; appendix table A7

[a] See supplemental materials for detailed annualized cost estimation. We use currency exchange of 1.1 USD = 1 EUR based on 2016 rates.

From Kittner, N., Gheewala, & Kammen, D. M., (2016). *Environmental Research Letters*, 11, 104013, Fig. 1, "Total cost estimates of electricity supply options."

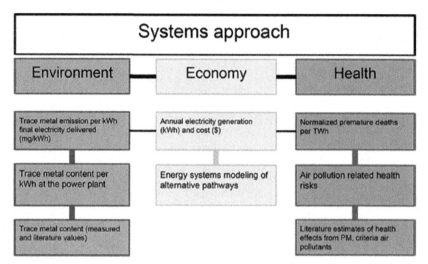

FIGURE 6.2 Overall systems approach. *From Kittner, N., et al. (2018). Trace metal content of coal exacerbates air-pollution-related health risks: the case of lignite coal in Kosovo.* Environmental Science & Technology, 52 (4), 2359–2367. *https://doi.org/10.1021/acs.est.7b04254, Fig. 1 "Overall Systems Approach".*

connection to health costs could not only benefit the study of Kosovo but also serve as an assessment model worthy of wider adoption. Moreover, the aerosolized states of these trace metals are difficult to quantify, particularly in regions lacking adequate air monitoring and sensing equipment. Categorizing the number of metals in the solid samples and then extrapolating their percentages in emissions would give researchers and decision-makers a new tool for assessing coal combustion in specific geographic areas.

Previous studies in the United States and Brazil had identified seven hazardous metals in various types of coal: arsenic, cadmium, chromium, mercury, nickel, selenium, and lead.

Coal-fired plants in the U.S. contribute approximately 62% of arsenic, 50% of mercury, 28% of nickel, and 22% of chromium emissions.

Arsenic is persistent, bioaccumulative, and toxic (PBT). It is a carcinogen and is particularly linked to lung cancer, skin disorders, and reduction of immune system function. Mercury impairs the nervous system and causes reproductive and developmental harm; increases the risk of certain cancers. Nickel is a carcinogen and is linked to chronic bronchitis, reduced lung function, dermatitis, and cancer of the lung and nasal sinus. Chromium is a carcinogen and has been linked to lung cancer, stomach ulcers, respiratory irritation, skin irritation, and sensitization (Table 6.3).

Chemists Heather Buckley, Raj Fadadu, and Zac Mathe, analyzed lignite coal samples from the main mine in Obilic, Kosovo. Buckley, Fadadu, and Mathe used inductively coupled plasma mass spectrometry (ICP-MS) to

TABLE 6.3 Trace metals present in lignite coals and their associated environmental and health impacts.

From Kittner, N., et al. (2018). Trace metal content of coal exacerbates air-pollution-related health risks: the case of lignite coal in Kosovo. *Environmental Science & Technology, 52* (4), 2359–2367. https://doi.org/10.1021/acs.est.7b04254, Table 1 "Trace metals present in lignite coals and their associated environmental and health impacts."

quantify the average mass of arsenic, mercury, nickel, and chromium that would likely be released from combusted lignite coal in Kosovo. Trace metals analysis was conducted by the Curtis and Tompkins Laboratory in Berkeley, CA, USA, according to EPA methods 3052, 6020, 7471A, for aluminum, arsenic, beryllium, cadmium, chromium, copper, lead, nickel selenium, silver, thallium, and zinc and mercury. The group found "... significant trace metal content normalized per kWh of final electricity delivered (As (22.3 ± 1.7), Cr (44.1 ± 3.5), Hg (0.08 ± 0.010), and Ni (19.7 ± 1.7) mg/kWhe)", posing "... health hazards that persist even with improved grid efficiency" (Table 6.4).

After collecting data on the amounts of trace metals in the samples, the team calculated an "emission factor"- how much mass of the metal in milligrams combusted in the coal-fired plant could be estimated to end up in air emissions and how would that compare to the amount of electricity that was generated? A spreadsheet model tested the trace metals amounts in different transmission, distribution, power plant efficiencies, and heat rates. The factor also took into account the known caloric values of different coal types. Using trace metal mass balances and chemical characteristics and literature estimates, the team estimated that 1%—10% of the As, Cr, and Ni mass would appear in the air emissions. The more volatile mercury was estimated to vaporize by as much as 80%.

In a separate but complementary analysis, public health researchers Meg Schwarzman MD and Dr. Tom McKone worked with the group to use this mg/kWh metric to calculate the likely morbidity and mortality rates based on standard PM criteria air emissions for the different energy production scenarios. The clear trend was that continued use of lignite coal in "business-as-usual" energy policy would likely generate 2200 greater deaths and thousands of incidences of illness among the Kosovo citizenry, as would be expected from an energy scenario employing solar and natural gas sources. This quantitative modeling data significantly strengthened the argument in favor of alternative and renewable energy sources and was published in the peer-reviewed journal "Environmental Science & Technology" in 2018 (Kittner et al., 2018).

Schwarzman and McKone were able to compare the four scenarios outlined by Kittner et al. (2016a,b) to establish an environmental and public health risk assessment for each. The options were: (1) constructing a new lignite plant, (2) using energy efficiency measures to meet Euro2030 targets, (3) transitioning to low-cost solar without natural gas backup, and (4) using solar augmented by natural gas for system flexibility. Annual energy generation portfolio values in kWh were entered into a software methodology called ExternE: Externalities of Energy. This model accounts for the reduction in life expectancy and cancers, e.g., premature death. The premature death endpoint estimates excess mortality attributable to exposure to PM2.5, sulfur dioxides, nitrogen oxides, and ozone (Table 6.5).

TABLE 6.4 ICP-MS heavy metal content in Kosovo lignite compared to lignite from other regions.

Heavy metal (CAS no.)[32]	Arsenic (7440-38-2)	Chromium metal, Chromium(II), Chromium(III), Chromium(VI) (7440-47-3)	Mercury (7439-97-6)	Nickel (7440-02-0)
Environmental Impact	Contaminates groundwater, disrupts plant growth and development, and decreases crop yields	Increases uric acid concentration in birds' blood and alters animal growth	Impairs nervous system and other organ systems in animals	Causes genetic alterations in fish and possible death and toxic to development organisms
Human Health Impact	Impairs immune system and increasing susceptibility to lung cancer	Causes reproductive and developmental harm and increases risk of certain cancers	Causes cognitive impairment in children and overstimulates central nervous system	Increases risk of lung cancer and causes nickel dermatitis
Boiling Point (°C)	465	2482	357	2730
Solubility in water (g/L)	20 (arsenic trioxide) at 20 °C	1680 (chromium trioxide) at 25 °C	74 (Mercury II chloride) at 25 °C	553 (nickel chloride) at 20 °C

From Kittner, N., et al. (2018). Trace metal content of coal exacerbates air-pollution-related health risks: the case of lignite coal in Kosovo. *Environmental Science & Technology*, 52 (4), 2359−2367. https://doi.org/10.1021/acs.est.7b04254, Table 2, "ICP-MS heavy metal content in Kosovo lignite compared to lignite from other regions."

TABLE 6.5 Air-pollution-attributable morbidity and mortality in four energy scenarios evaluated in Kosovo's power sector projected for 2016—2030.

Scenario	Deaths	Air-pollution-related risk	
		Serious illness	Minor illness
Business-as-usual	3200 (800—12 700)	29 000 (7300—88 000)	1 700 000 (430 000—6 900 000)
Euro2030	2000 (510—8100)	18 500 (4600—75 000)	1 100 000 (280 000—4 400 000)
Solar without natural gas	1300 (320—5200)	12 000 (2900—47 000)	700 000 (180 000—2 800 000)
Solar with natural gas	900 (230—3600)	8400 (2100—33 700)	460 000 (120 000—1 800 000)

From Kittner, N., et al. (2018). Trace metal content of coal exacerbates air-pollution-related health risks: the case of lignite coal in Kosovo. *Environmental Science & Technology*, 52 (4), 2359—2367. https://doi.org/10.1021/acs.est.7b04254, Table 3 "Air-pollution-attributable morbidity and mortality in four energy scenarios evaluated in Kosovo's power sector projected for 2016—2030."

3. A campaign of education and advocacy

Data alone is not sufficient for shifting policy, and technical reports rarely permeate the news media ecosystem with the author's original intent intact, so Kittner and Kammen wrote and presented extensively to communicate their findings to nontechnical audiences. This vigorous science communication effort gave policymakers, community leaders, NGOs, and everyday Kosovars additional information about the benefits and feasibility of adopting a diverse renewable energy infrastructure. These advocacy articles were published in multiple publications including The Economist, The Beam, the See Change Network blog, The Nikkei Asian Review, the San Francisco Chronicle, and various publications by KOSID. Kittner also presented the model at conferences and fora worldwide.

These engagement efforts were vital for garnering popular support for pursuing diverse renewables in Kosovars' journey toward energy independence. For example, a robust renewables market would allow Kosovo to buy less energy from Serbia, a rival state that refuses to recognize Kosovo as an independent nation. This, in turn, would give Kosovo much-needed political flexibility and create substantial economic opportunities for local green energy jobs in Kosovo. Kittner and Kammen argued further that boosting Kosovo's green energy sector would align it more with European Union (EU) membership requirements, bolstering the country's lagging bid to join the EU.

What was innovative about the solution

Developing an effective renewable energy solution for Kosovo required interdisciplinary coordination among experts in economics, energy, chemistry, and public health. It also required a robust affiliation with local public interest groups and officials in Kosovo, the European Union, and the World Bank. This was a novel approach that developed a metric for trace metal content per electricity output. This new tool took basic chemistry lab work results and translated them into a useable metric for use in the sphere of economics and financial decision-making. Further, the team was able to align these results with a more standard measurement of public health risks, calculate premature deaths and illnesses, and integrate this into a compelling overall economic case for renewable energy production in Kosovo.

Beyond the immediate challenge in Kosovo, these methods could serve as a model to more accurately assess public health costs in another development project across the globe. Similarly, and most significantly, the framework provides a roadmap for diverse and flexible renewable energy sources that can respond to spikes in electricity demand more effectively and create a more durable, robust grid system.

What was the impact

The energy production and distribution models developed for this project have been published and made available for public use, as has the paper on mg/kWh, the novel metric for measuring trace metal pollution per kilowatt hour (kW) hour of electricity produced. Thanks to support from BCGC, Noah Kittner published five peer-reviewed papers, over a dozen science communication articles, and traveled to Kosovo to work directly with KOSID, but the most significant impact of this project was an announcement from the World Bank regarding energy infrastructure investment in Kosovo.

On October 10, 2018, The World Bank announced that it would not be financing a new coal plant in Kosovo, because alternative energy cost assessment models indicated that renewable energy sources were cheaper than conventional coal plants, particularly when the economic cost of health impacts from lignite coal production was considered. On hand to help announce the World Bank's position was a coauthor of the study that Kittner and Kammen produced in collaboration with KOSID, proving the power of collaboration between academic research groups and local partners. The scale and magnitude of the outcome of this project were not only due to the sophistication of the research but also the deliberate efforts to connect academic research with stakeholders and policymakers.

Despite this loss of powerful guarantees from the World Bank that would unlock other loans, the government of Kosovo decided to continue with their reliance on coal, citing the greater security of using local resources, existing infrastructure, and the lack of their own development capital. Kosovo has 14 billion tons of lignite, the world's fifth largest deposit. In November 2018, they contracted with the London-listed ContourGlobal to construct and operate a new 500 MW coal-fired plant and agreed to buy all of its output for 20 years. Since that time, ContourGlobal has partnered with U.S. giant General Electric to build the plant, slated to start in 2019.

At the same time, Kosovo's largest beer maker, Peja Beer, has constructed a 6 MW solar park outside the western town of Gjakova, producing 3.6 MW h of electricity. While renewable energy expansion is part of the government's policy, production is still relatively small, but a strong debate about its energy future and its public health continues within the country.

The final outcomes for Kosovo remain to be seen, but this case study remains a testament to how collaborative science can inform powerful decision-makers and serve as a model for wiser stewardship of our resources and our public health.

References

2030 EU climate and energy framework outlined in the 2015 Industrial Emissions Directive. Available at: https://ec.europa.eu/clima/policies/strategies/2030_en.

Kittner, N., Dimco, H., Azemi, V., Tairyan, E., & Kammen, D. M. (2016a). An analytic framework to assess future electricity options in Kosovo. *Environmental Research Letters, 11*(10), 104013.

Kittner, N., Fadadu, R. P., Buckley, H. L., Schwarzman, M. R., & Kammen, D. M. (2018). Trace metal content of coal exacerbates air-pollution-related health risks: The case of lignite coal in Kosovo. *Environmental Science & Technology, 52*(4), 2359−2367.

Kittner, N., Gheewala, S. H., & Kammen, D. M. (2016b). Energy return on investment (EROI) of mini-hydro and solar PV systems designed for a mini-grid. *Renewable Energy, 99*, 410−419.

World Bank October 10, 2018 Announcement. Available at: https://openknowledge.worldbank. org/bitstream/handle/10986/13216/750290ESW0P1310LIC00Kosovo0CEA0Rprt.pdf? sequence=1.

World Bank. (2013). *Toward a sustainable energy future for all:directions for the World Bank Group's energy sector Washington DC*. World Bank. Available at: http://documents. worldbank.org/curated/en/2013/07/18016002/towardsustainable-energy-future-all-directions-world-bankgroups-energy-sector.

Further reading

Atkin, E. (Jun 29, 2015). The supreme court just delivered A victory to coal plants that want to emit unlimited mercury. *Think Progress*. Available at: https://archive.thinkprogress.org/the-supreme-court-just-delivered-a-victory-to-coal-plants-that-want-to-emit-unlimited-mercury-1a4ebea121ba/.

ATSDR. https://www.atsdr.cdc.gov/phs/phs.asp?id=243&tid=44.

ATSDR. https://www.atsdr.cdc.gov/toxfaqs/tf.asp?id=47&tid=15.

ENV. https://www.env-health.org/wp-content/uploads/2018/12/HEAL-Lignite-Briefing-en_web. pdf.

EPA. https://www.epa.gov/lead.

EPA. https://www.epa.gov/sites/production/files/2016-10/documents/hydrogen-fluoride.pdf.

Europa. https://ec.europa.eu/energy/home_en.

Facebook. https://www.facebook.com/KOSID.Kosovo/.

Fadadu, R., Kittner, N., & Kammen, D. (November 2019). Air pollution-related health risks associated with energy production in Kosovo. In *APHA's 2019 annual meeting and expo (nov. 2-Nov. 6)*. American Public Health Association.

Hooper, E., & Medvedev, A. (2009). Electrifying integration: Electricity production and the South East Europe regional energy market. *Utilities Policy, 17*(1), 24−33.

Kittner, N., Dimco, H., Azemi, V., Tairyan, E., & Kammen, D. M. (2011). *Sustainable electricity options for Kosovo*.

Kittner, N., Dimco, H., Azemi, V., Tairyan, E., & Kammen, D. M. (2014). *Sustainable energy pathways for Kosovo. Technical report* (p. 2014). Kosovo Civil Society Consortium for Sustainable Development.

Kittner, N., Fadadu, R. P., Buckley, H. L., Schwarzman, M. R., & Kammen, D. M. *Supporting information for: Trace metal content of coal exacerbates air pollution-related health risks: The case of lignite coal in Kosovo*.

Kittner, N., Lill, F., & Kammen, D. M. (2017). Energy storage deployment and innovation for the clean energy transition. *Nature Energy, 2*(9), 1−6.

Klubiprodhuesve. https://klubiprodhuesve.org/en/devolli-group-member-of-manufacturing-club-installed-the-largest-photo-voltaic-park-in-the-region/.

Krieger, H. (Ed.). (2001). *The Kosovo conflict and international law: An analytical documentation 1974—1999*. Cambridge University Press.

Linkedin. https://www.linkedin.com/company/curtis-&-tompkins-laboratories/about/.

Monographs. https://monographs.iarc.who.int/wp-content/uploads/2018/06/mono87-1.pdf.

Reuters. https://www.reuters.com/article/us-contourglobal-kosovo-ge/ge-to-build-kosovos-new-500-mw-coal-power-plant-idUSKCN1S917R.

Reuters. https://www.reuters.com/article/us-kosovo-energy/kosovo-opts-for-coal-plant-despite-criticism-idUSKCN1N71LM.

Surgery. https://surgery.duke.edu/news/despite-studies-health-effects-coal-burning-power-plants-remain-unknown.

Toxtown. https://toxtown.nlm.nih.gov/sources-of-exposure/power-plants.

UNI. https://uni-pr.edu/.

USE, P. (1996). *Method 3052: Microwave assisted acid digestion of siliceous and organically based matrixes*. Washington, DC USA: United States Environmental Protection Agency.

USE, P. (1998a). *Method 6020a: Inductively coupled plasma—mass spectrometry*. Washington, DC USA: United States Environmental Protection Agency.

USE, P. (1998b). *Method 7471B (SW-846): Mercury in solid or semisolid wastes (manual cold-vapor techniques)*. Washington, DC USA: United States Environmental Protection Agency.

Chapter 7

Conclusions

In this book, we have attempted to place Green Chemistry within the context of sustainability and the great existential issues of our time: climate change, growing scarcity of resources for a growing population, and the continued degradation of the health of our environment. We have shown that the practice exists within a matrix of policy and law, as well as the interwoven interests of the commercial, governmental, and societal. We have laid out a process for the investigation of substances of concern, and how safer alternatives can be assessed, and how innovative solutions might be found. Four case studies in textiles, household products, additive manufacturing, and renewable energy detail this process and the rarely predictable outcomes from what is typically a multi-actor, multi-phase, and trans-disciplinary venture.

Despite the varied subjects of these investigations, several common themes have emerged over the course of 9 years of teaching the Greener Solutions graduate course at UC Berkeley: complex problems are rarely solved with simple solutions; they require integrated intelligence across disciplines and this is usually best achieved with a model of collegial sharing of deep subject matter within a team of experts. How this intelligence is shared and the way that team members think and interact with each other is often as important or more important than what they know. Everyone has something to contribute, and there is always, always, the unexpected. Often this leads to real progress and sometimes to innovation.

We hope that this text can be used as a guide for students and practitioners, and for educators who would like to establish a similar course or module at their institution. Further, we hope that the model of academic/industry/government partnership at the Berkeley Center for Green Chemistry will be taken up by others and lead, in other ways and through different paths, to safer chemical design, manufacturing, and use.

Green Chemistry in Practice. https://doi.org/10.1016/B978-0-12-819674-8.00009-6

Index

Note: 'Page numbers followed by "*f*" indicate figures and "*t*" indicate tables.'

Printed in the United States
by Baker & Taylor Publisher Services